DATE DUE

Satellite Communications in the 21st Century: Trends and Technologies

Satellite Communications in the 21st Century: Trends and Technologies

Edited by

Takashi Iida
Communications Research Laboratory
Tokyo, Japan

Joseph N. Pelton
George Washington University
Washington, D.C.

Edward Ashford
SES GLOBAL
Beltzdorf, Luxembourg

Volume 202
PROGRESS IN ASTRONAUTICS AND AERONAUTICS

Paul Zarchan, Editor-in-Chief
MIT Lincoln Laboratory
Lexington, Massachusetts

Published by the
American Institute of Aeronautics and Astronautics, Inc.
1801 Alexander Bell Drive, Reston, VA 20191-4344

ISBN 1-56347-579-0

Progress in Astronautics and Aeronautics

Editor-in-Chief

Table of Contents

Preface

The editors and contributors to this book wish to thank AIAA for providing the opportunity to publish. It is a somewhat unusual book, in that it examines not only the past and the evolution of satellite communications up to the present time, but it also analyzes the likely future of satellite communications. Advanced digital satellite technology is shaping our world in important ways today, and the new capabilities and research directions analyzed and addressed in detail in this book shine a light into the incredible next few decades of the space industry.

This book was designed for several different types of readers, including both students interested in satellite technology, whether studying in a technical field or not, as well as those active today in the satellite industry. The first two chapters allow those who may not already be familiar with communication satellites to understand the basics of the industry, the fundamentals of the market, the applications that exist, and especially how the technology works. Later chapters go into detail in specific communication satellite application areas.

An ancient Greek, even prior to the age of Socrates and Plato, is reputed to have said that, "The more things change, the more they remain the same." This seems to us particularly true as well for the evolution of satellite communications technology and systems. The communication satellite industry has been replete with change from the very beginning, as new power systems, new stabilization and pointing capabilities, new and more sophisticated antennae configurations, smaller and more affordable user terminals, and the use of new frequency bands and new orbits have all evolved since the mid 1960s. However, this constant pattern of change has allowed the satellite industry to remain constant, in that it allowed it to keep up with new applications and services and to meet the growing worldwide demand for advanced digital telecommunications services.

Satellite technology must continue to develop over the coming decades if the satellite industry is to remain constant in its ability to provide affordable services. Furthermore, only through continued development will the industry "continue" to be competitive with competing telecommunications technologies such as fiber optics and broadband terrestrial wireless.

Fortunately a dynamic research and development process is in being in many space-faring nations to ensure the future growth, development, and prosperity of the satellite industry. This is supporting the ongoing technological change that is needed to meet the same and ongoing demands of the growing population on our small planet. There is an old adage that says, "You cannot miss that which you have never experienced." There is a corollary to this as well, however, to the effect that, "Once you have experienced something useful or enjoyable, you always want more." The majority of the more than six billion people now on Earth have never experienced the benefits of modern communications. Indeed, most have never even had the opportunity to even make a telephone call. This is changing, however, and communications technology is becoming pervasive, even in remote and under-developed areas. By the time, predicted to be within some 30 years, that Earth's population grows to over 8 billion, a far higher percentage of these will have experienced the benefits of modern telecommunications, and will demand more and more from it. Constant change in satellite telecommunications technology is most certainly needed to meet these expanding global communications needs—in commerce, entertainment, health and education, peaceful interaction, and conflict resolution. Satellite technology must continue to provide systems that extend and improve safety and global reach, that are compact, low cost and "user friendly," that can provide broadband services to the most remote locations, that enhance mobility, and that provide affordable Internet access.

From the outset, the satellite industry has been international in scope, with the Intelsat 1 (or Early Bird) satellite proving in 1965 just how valuable satellite networks can be in increasing global communications. Since that time, satellites have been employed to create improved patterns of world trade, to facilitate global news exchange, and to engender international understanding and intercultural exchange through, for example, satellite tele-health and tele-education programs and other forms of global networking. For those who doubt that satellite technology will be able to keep pace in the 21st century, this book explains not only today's capabilities but seeks to direct a technological spotlight into the future. We hope that this spotlight will prove illuminating as well as thought provoking to its readers.

Takashi Iida
Joe Pelton
Ed Ashford
July 2003

Acknowledgments

Chapter 1

I would like to thank all the authors of this book who have offered advice and recommended changes to this chapter. Special thanks must go to Takashi Iida, who first conceived of this book, and to Ed Ashford, who has given much time and attention to editing this chapter and the entire book to greatly improve its quality.

Chapter 2

A chapter introducing the principles of satellite communications includes, by its very nature, little original information. I have attempted to take the work of many others that have contributed to the field and summarize and condense information from them into a form that would be understandable to the intended audience. A number of people assisted me in this by agreeing to let me use excerpts or diagrams from their previous publications. In this regard, I am grateful in particular to Mark Williamson (who made some valuable comments to an early draft of this chapter), Simon Dinwiddy (of ESA/ESTEC in the Netherlands) and Ruud Weijermars (Project Manager of SpaceTech studies at the Technical University of Delft, also in the Netherlands).

I would also like to thank Tom Mannes and Bruno Perrot who improved early drafts, and especially Ray Sperber of SES ASTRA for reviewing the "almost final" draft of this chapter. Their suggestions for additions and changes have eliminated a number of my errors and helped greatly to make the text more accurate. Any remaining mistakes and, in particular, any that may have crept in after Ray's review, remain my responsibility alone.

I would also like to thank both my wife, Mercy, and daughter, Stephanie, for their help in reading this chapter and correcting my instances of misspelled words and poor grammar. Finally, I would like to thank Mercy again for putting up with my absence during many evenings and weekends while I did research, writing, or editing of this and other chapters.

Chapter 5

I would like to thank Detlef Schulz of SES ASTRA for his time and patience in explaining the history and capabilities of the DVB standards.

Chapter 6

The authors would like to give their thanks to R. Miura, Y. Arimoto, and R. Suzuki for their help to in organizing this chapter.

Chapter 7

The authors would like to give special thanks to the late Burton I. Edelson, as well as Neil Helm and Ivan Bekey of the George Washington University for their suggestions and participation throughout the study on the GEO platform. The authors also would like to thank Akira Akaishi and Sachio Shimoseko of Mitsubishi Electric Corporation, and Yoshiaki Suzuki of CRL for their help in clearly presenting the 2G-satellite design concepts.

Chapter 1

Overview of Satellite Communications

Joseph N. Pelton*
George Washington University, Washington, D.C.

I. Introduction

COMMUNICATIONS satellites have been changing our world for more than 40 years. Further, there is an excellent chance that satellites—although exotic in form and function yet still satellites—will continue to provide us with change and innovation for at least another 40 years. Those who want to explore the potential of these satellites at the beginning of the new millennium have an interesting read ahead of them as we explore the exciting new technologies that will soon move out of the laboratory and into operation to provide the world with exciting new services.

The prospects for new technology and systems that will come in future decades are nothing short of astonishing. Michael Nelson, a former leading U.S. telecommunications policy maker and strategic planner for the IBM Corporation, recently observed that within the next 20 years there will be at least a thousand times more connections on the Internet and corporate Intranets than there are today, while the information stored on the Internet may increase by more than three orders of magnitude. Nelson further predicted that the number of people around the world who would be electronically networked together via fiber, wireless, and satellite would increase by a factor of at least four to five times. In a short period of time, phrases such as "the global brain," the "worldwide mind" or the "e-sphere" will take on true meaning and redefine patterns of learning, work, and leisure in an ever-expanding global culture. Satellites will be a meaningful part of this future.

If Nelson and other observers such as Bill Gates, Nicholas Negroponte, and Secretary General Yoshio Utsumi are correct, telecommunications and networking markets have only begun to define a small portion of the new space services that will abound in the 21st century. Satellites have become so prevalent and widely used that it is easy to overlook the remarkable ways they have already instilled change into an increasingly global society. In fact, satellites have actually served as a prime enabling technology to allow true modern globalism to exist. While

---

*Director, Space and Advanced Communications Research Institute.

fiber-optic cables have woven together patterns of knowledge in the East to West directions of the global society, satellite systems have increasingly served to bring together North to South and urban to rural.

II. Satellites on the World Stage

Our amazing new space systems have brought us global television and radio on a previously unimagined scale, with over 10,000 satellite television channels now in operation. Billions of people can now watch worldwide news, sports, and entertainment 24 hours per day, 365 days per year. These global satellite audiences are now 10 to 100 times larger than any audience ever achieved in the presatellite days. There are now more than 200 countries and territories linked together via telephone, fax, telex, data, and now perhaps most interactively, via the Internet. Satellite systems link us in ways that are more immediate, interactive, intercultural, and international than land lines, cables, or terrestrial wireless systems.

A. Expanding Global Satellite Markets

Today, e-commerce, Intranets, and the Internet reach even the most remote portions of the Earth, in South and Central America, the Caribbean, the Asia-Pacific, the Middle East, and Africa. Indeed, many of these regions can still be reached only by satellite. The chart in Fig. 1.1 shows three different consulting firms' recent projections as to e-commerce growth on a global scale.

Today, approximately three-quarters of the Earth's surface (oceans, deserts, rain forests, mountain ranges, swamps, and bogs) can only be instantly connected by satellite. An even larger portion of the world's airspace can be reliably and consistently linked only by satellite systems.

Satellite services have grown up over time and were, as of year-end 2002, a $48 billion per year global industry when all retail and value-added services are considered. Table 1.1 shows Merrill Lynch's financial analysis[1] of past and projected trends in the global growth of overall revenues in the satellite field. The

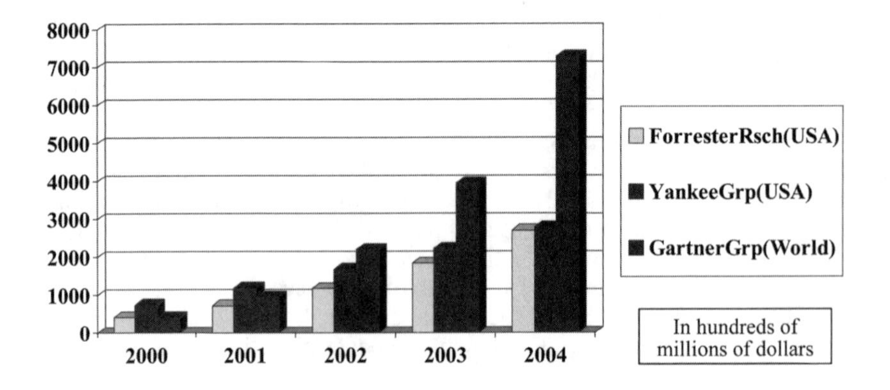

Fig. 1.1 Global growth of e-commerce.

Table 1.1 Global growth in total satellite communications revenues (2000–2007) (Ref. 1)

Year	Total revenues/year ($billion)
2000	$30 billion/year
2002	$48 billion/year
2003	$59 billion/year
2005	$83 billion/year
2007	$113 billion/year

main uncertainty with regard to these projected figures is in the estimates for mobile satellite and broadband satellite services. Some analysts believe these figures will rapidly increase in the 2003 to 2007 time period, as indicated by the Merrill Lynch and Gartner Group Data Quest projections; others such as Euroconsult, the International Engineering Consortium, and DTT Consulting believe that growth in these areas will be more modest.

Regardless of whether one is bullish (total revenues up by a factor of 2.5 by year-end 2007) or bearish (total revenues up by a factor of 2.0 by year-end 2007) some measure of significant and sustained growth is still very likely to be the case.

There is solid evidence that satellite services will continue to grow even with the spread of potentially competing fiber-optic cables and terrestrial wireless services. This is because most communications and IT networks are not represented as "zero-sum games" when it comes to space and terrestrial technologies. In ways that will be explored later in this book, satellite and ground systems will, perhaps more often than not, grow in tandem by offering complementary services. Satellites and fiber systems often play to each other's strengths and supplement each other's weaknesses.

Direct broadcast satellite services have shown continued growth of more than 20% per annum for a number of years, although sales started to decline in the early part of 2002. Broadband satellite services, although starting from a relatively small base of a few hundred thousand subscribers in the U.S. and Europe, were projected to grow by 75% in 2002. Even traditional fixed-satellite services, led by Internet-related services, are expected to grow by 7 to 8% in the years immediately ahead.[2,3]

Although mobile satellite services and other messaging services experienced limited and somewhat unpredictable market growth when they were first introduced, the overall pattern for the satellite industry, starting in early 2002, was a recovery from the market setbacks experienced because of global and U.S. recession-related trends that first became apparent with the crash of the "dot.com bubble" in early 2000.

This longer-term projection of continued satellite market growth is based on the belief that space communications systems are uniquely positioned to provide certain special services and applications. Additional new satellite technology and continued innovations with respect to user microterminals will allow better, more reliable, and lower-cost space telecommunications and provide new forms of digitally integrated services. Furthermore, there are some social, economic, scientific, and governmental services that involve broadcasting or multicasting that best lend themselves to satellite networks. Finally, in rural and remote areas, two-way satellite connections will provide the broadband access that would otherwise be unattainable by other means.[4]

B. Historical Patterns of Satellite Growth

When commercial satellite services began in 1965 with the Early Bird satellite, options were limited and at considerable expense. Initially, there was huge demand for high-quality international communications, and satellite system engineers believed that their main challenge was to develop and launch ever-larger and higher-capacity systems as quickly as possible. In those early days, new satellite generations were deployed every 2 to 3 years, with capacity increases as large as five times that of the previous satellite designs. Today, the markets are better established and satellite designs are increasingly being geared to meet the next type of service demand and are thus largely governed by conservative projections of market demand. However, we have seen exceptions to this rule with the Iridium, Globalstar, ICO, Astrolink, and Orbcomm satellite systems during the period of 2000 to 2002.[4]

The size and performance of satellite systems continues to grow quickly. There is a continuously growing array of new space communications and information services that new satellite designs must take into account. These include space navigation, fixed satellite services, voice-over IP, digital video broadcast with return channel service, broadband Internet, and IP networking and multicasting services including cache updating, multimedia, and video streaming. Further, there is a rapidly growing demand for digital television; high definition television (HDTV) services; direct access radio services (DARS); mobile satellite services for aeronautical, maritime, and land mobile connections including 3G broadband mobile services, messaging, and paging; Supervisory Control and Data Acquisition (SCADA); intersatellite links; meteorological and remote sensing data relay; radio distance determination; and geodetic, scientific, and time measurement services. The list could include more than one hundred different offerings. Not all of the manifold types of services can be addressed in this book, of course. What will be addressed, however, are the various types of generic telecommunications services—fixed, mobile, and broadcasting—into which this wide array of services can be classified.

III. The Evolution of New Satellite Designs to Meet Specialized Service Demand

Since the beginning of commercial services in 1965, specialized satellites with different power levels, frequency bands, antenna coverage patterns, orbital configurations, and onboard intelligence and processing capabilities have evolved to respond to all of the diverse new markets previously mentioned. Critical to this process has been the development of different types of user terminals with a range of performance capabilities, in a variety of sizes, and with dramatically different costs. In less than four decades, satellite Earth stations have gone from those that were 30 m in diameter, cost millions of dollars, and required 30 to 50 staff members to operate, to today's desktop microterminals and handheld satellite transceivers. Between these extremes, a wide range of different Earth station types and sizes coexist in today's diverse market.

Satellites therefore now service an ever-widening array of customers. These include business users, small office and home office users, ships at sea, military and civilian aircraft, scientists on land and sea, and consumers receiving home entertainment or obtaining broadband Internet services. All of these users and more have increasingly demanded specialized satellite services. Likewise, they have also sought to obtain new and optimized user terminals designed to meet their own needs.

The future growth of satellite services will be governed by new technology; changing market demand; new standards; new legal, policy, and regulatory systems; and competitive pressures in the marketplace. Of these factors, market demand is likely to be the most important influence on the future of satellites. These demands will come from business-to-business applications (B2B); entertainment; governmental, social, and scientific applications; and mass consumer demand. More precisely, the future of satellite growth will be stimulated primarily by the following factors:

1) The global demand for broadband connectivity spawned by the Internet and e-commerce. This will be complemented by the demand for wide-area broadcast entertainment, interactive broadcast, and multicasting applications to support consumer, business, science, education, and health applications.
2) New telecommunications standards that will allow seamless integration of all types of digital services.
3) The further opening of international trade in international communications services as stimulated by the World Trade Organization (WTO) and General Assignment on Trade in Telecommunications Services (GATS) and the increased sway of regional trade groups such as the Asia-Pacific Economic Community (APEC), the North American Free Trade Agreement (NAFTA), and the European Union (EU) among others.
4) The development of new, improved, and more cost-effective satellites and user terminals to serve both fixed and mobile applications and also increasingly provide direct mass-market access to customers plus facilitate effective B2B applications in a user-friendly mode.
5) Increased deregulation, liberalization, and competition in all forms of telecommunications services coupled with the competitive effect of new fiber-optic cable and broadband wireless services on the demand for more cost-effective and powerful satellite performance.
6) New developments that allow satellites to provide increased security, reliability, and flexibility within digital IP networks while also supporting new and ever more directly interlinked "MESH" architectures.

The trend, which began in the early 1990s and has moved the world economy toward deregulation, the breakdown of telecommunications and IT monopolies, and the onset of intense international service competition, has brought about changes at every level in the field of telecommunications. We have seen increased levels of innovation, a sharp rise in digital convergence, expanded global reach, and new demand for broadband services, coupled with a need for flexibility and a diversity of value-added services.

The last two decades have seen the evolution of a new competitive marketplace. This has been coupled with the demand for fair access, both to the international business community and to consumers. These changes toward competition, liberalization, and deregulation are the essence of modern telecommunications systems. This is especially true for satellite networks defining a new role for themselves in this new competitive environment.

Fiber systems can provide virtually all of the desired characteristics of this new competitive telecommunications marketplace, particularly very low-cost services. Figure 1.2 shows that the cost of a fiber-based telephone channel has now dropped to the level of \$5 to \$10 per year level and is still dropping. However, new connections via fiber systems can take months or even years to install, while new satellite links can be activated in minutes. Satellites can add truly global coverage to even the most remote areas and can also add mobility and access to networks of almost infinite size. Moreover, on the most advanced satellites such as the Intelsat 9, the annual capital cost of an 8 kbps voice circuit is also now less than \$50 per

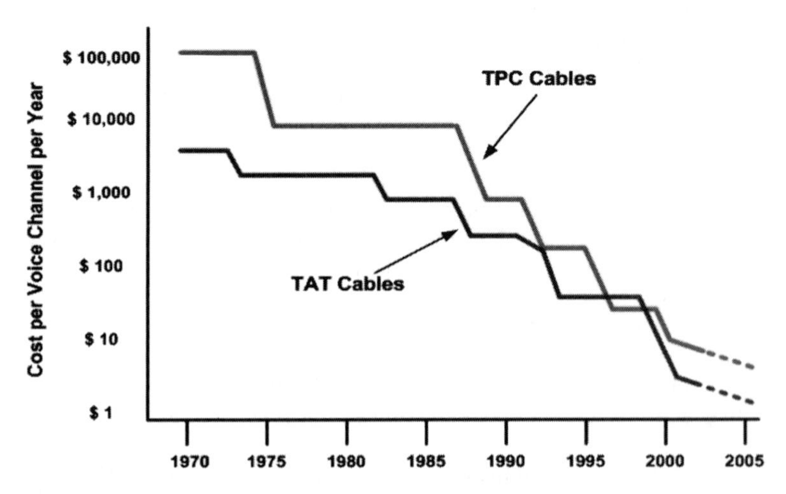

Fig. 1.2 Dramatic decrease in the capital cost of a voice channel on a fiber-optic cable. (This chart was compiled by J. N. Pelton from data that was derived from *The New Satellite Industry*, Telegeography Inc., 2001 and other reports.)

year. Thus, the need to use satellites and terrestrial technology in a balanced and complementary way seems both prudent and actually largely unavoidable.

The purpose of this book is to examine changing trends in satellite markets and in communications satellite and user terminal technologies. The purpose throughout is not only to interpret the past, but also to project, hopefully with some accuracy and reasonable factual foundation, the future course of satellite communications.

Some technologies are currently being adapted from the lab and implemented on new operational systems; thus, forecasting their evolution is quite straightforward. The longer-term future, however, is more complex and difficult to predict. Nevertheless, there are certain established trends that seem rather powerfully established. It seems unlikely that a single aspect of new satellite technology will dominate in the future, but rather there will be a continuing diversity of new technologies and services. New materials, new processing technologies, new launch services, and new concepts such as high-altitude platform systems will assist the process of creating new satellite cost efficiency and continuing this trend. Such new technologies will allow future satellite systems to provide faster throughput, more integrated services to smaller and lower-cost microterminals, and improved value-added services at lower cost. How this might happen will be explored in the chapters to follow.

Further, the world and the world of satellites will remain diverse. There will be different types of market success, and these will occur in different parts of the world and in different service sectors. Today, at least, there is no such thing as an integrated satellite market, either from a geographic or technological basis. Some types of markets such as fixed satellite services, direct broadcast entertainment, and interactive video game services are doing well, while others, such as land mobile and broadband satellite services, are greatly lagging behind earlier market projections. This pattern will change over time. The authors believe that the best days of mobile satellite services and of broadband satellite networking are still to come.

The evolution of new digital satellite technology, multibeam satellite technology, onboard signal processing and regeneration, advanced and more efficient power systems, and other gains could allow the formation of digitally integrated satellite markets via highly cost efficient multipurpose satellite systems within the next decade. In another decade we may even see totally new types of satellite technology and architecture, which could allow satellites to perform a range of new services and achieve dramatic new economies of service. The future of satellites is thus exciting, evolving, and in many ways just beginning.

Therefore, this book on the future of satellites will include different analyses with respect to Internet and IP services, mobile satellite services, broadcast satellite services, the next generation of satellites, and an even longer-term analysis of satellites systems of the next 30 years.

IV. Telecommunications and Satellite Megatrends

To understand satellite communications, one must understand the context of the overall emerging digital services on a worldwide basis. In short, satellites cannot be understood in a vacuum. There are clearly large-scale "Mega Trends" that shape the satellite world of the future.

A. Super Speed—The Intensity of Modern Technological Change

The high-tech world of computers, networking, robotics, communications, and remote sensing all, more or less, follow Moore's law. That is to say, there is a doubling of capacity every 12 to 18 months. Satellites are indeed a part of the computer world. Modern satellites are projected to increasingly become essentially very high-speed digital computer processors in the sky. This trend, at least in part, supports the trend to achieve not only speed but economies of scale associated with extremely high throughput.

The latest generation of Intelsat satellites, for example, Intelsat 9, contain some 96 transponders and are capable of prodigious throughputs. Although there are cases to be made in some instances for smaller spacecraft to serve specialized roles, the move to super speed broadband services tends to favor larger spacecraft. This has led to projections that there will be less launches (especially for geostationary satellites) in the future but that because of higher performance, frequency reuse, and power, these satellites will represent overall global growth. This projection is shown in Fig. 1.3.

B. Internet Growth and E-Services

The astonishing growth of the Internet continues, even with the so-called burst of the dot.com bubble in early 2000. Internet users worldwide topped 500 million

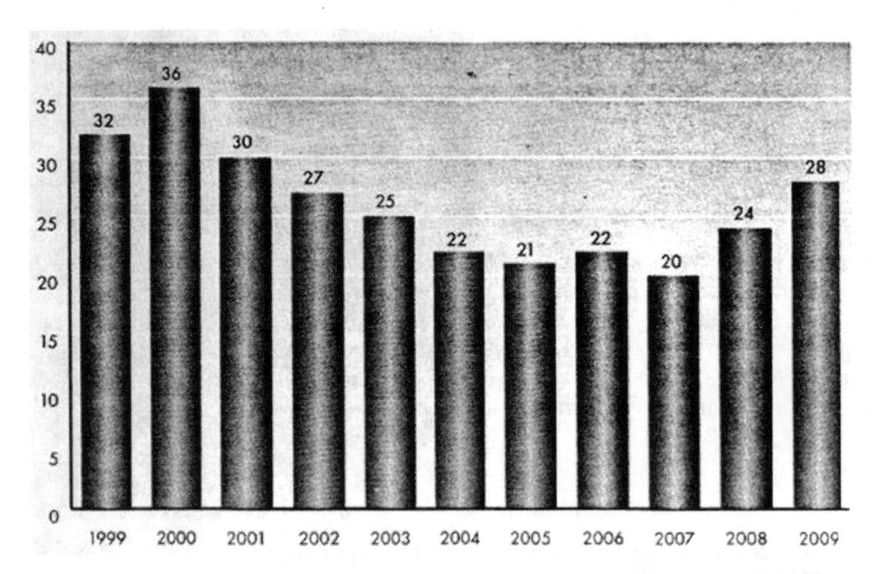

Fig. 1.3 Projected number of Geostationary satellite launches through 2009.

in 2001, which is one-half billion people around the world that are interconnected virtually instantaneously. This is expected to increase to more than one billion users by the end of 2005.*

According to the Gartner Group's Dataquest surveys,[2] it is expected that by the end of 2003 there will be some 350 million users in just North America and Europe, and another 180 billion in the Asia-Pacific. The percentage rate of increase with respect to the total population is now greatest in South and Central America, the Middle East, Oceania, and Africa. Direct satellite access and multicasting services will only speed the increase of Internet and e-commerce around the world. E-commerce has tapered off from its most explosive growth, but it is still expected to grow at more than 30% per annum for at least several years to come. Today, more than 20% of all satellite growth in volume and revenues is Internet and IP related in all areas, whether in B2B services, direct access, multicasting, video streaming, or entertainment.

As of April 2003 there were more than 63 million internet users via mobile phones in Japan.† Third-generation PCS services are now in full-scale implementation in Europe, although introduction is just beginning in the United States. There can be little doubt that Internet broadband services to wireless networks are the next wave of growth in telecommunications services. The real question is the extent to which wireless mobile Internet services will eventually stimulate the growth of mobile satellite services. The first wave of mobile satellite system growth, which was to be based primarily on voice, paging, and messaging services, did not succeed, and many of the early ventures such as Iridium and ICO and Globalstar experienced major market failure. The question is whether broadband Internet services can fuel success in coming years.

C. Business and Consumer Demand for Broadband Services

The demand for broadband services is the core growth market for telecommunications services for the current decade. A few years ago, some believed that the answer to providing broadband services was simply to provide fiber-optic access to the home and the desktop. The so-called Negroponte Flip, as set forth in 1993, projected that there would not be enough spectrum to meet the demand by wireless or satellite systems and thus all broadband would have to go by means of fiber-optic systems.[4] Negroponte suggested that all narrowband wireless would handle voice and messaging, but that everything else would migrate to cable-based systems (see Fig. 1.4). The process has turned out to be more complicated than expected. The initial demand for mobile wireless telephone and messaging services in the 1990s has turned into a torrent of new market demand for broadband mobile services as well. Thus, the so-called Pelton Merge also set forth in 1993 suggests that broadband integrated digital services will need to be seamlessly provided by an "electronically fused" network of wireless, satellite, coax, and fiber-optic systems. The demand for rural and remote services

*Data available online at http://www.c-i-a.com [cited March 2003].

†Data available online at http://www.soumu.go.jp/S-news/2003/030530-1.html [cited May 2003].

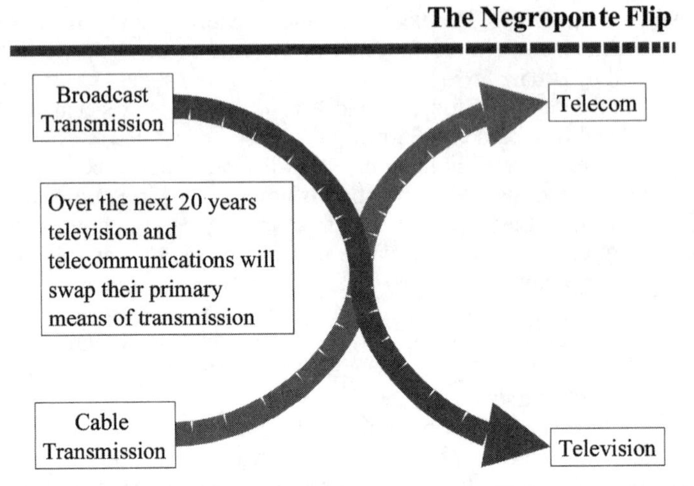

Fig. 1.4 The Negroponte Flip predicted all wideband services would go on fiber-optic cable but demand for mobile services has steered the market in new directions.

and broadband mobile simply does not allow a single media, such as fiber-optic cable, to provide a unified solution (see Fig. 1.5).

Today, there are indeed varieties of wire and wireless systems working together to provide broadband services. These include a maze of wireless technologies. There are wireless LANs (802.11), to 2nd and 3rd generation personal

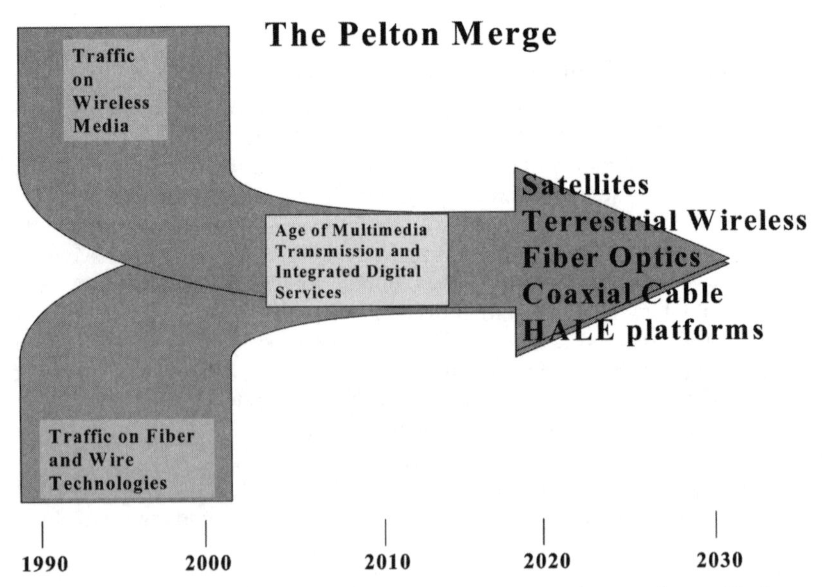

Fig. 1.5 The Pelton Merge predicted that diverse market needs would demand seamless digital standards for transmission media to work together.

communication systems (PCS), to MMDS and LMDS (microwave and local multipoint distribution services, respectively) wireless cable systems, to free-space infrared buses and laser systems, to various types of satellite direct access networks. All of these and more are serving the broadband mobile market.

This myriad of wireless networks also needs to be effectively linked to wire-based systems. Here, broadband services are being provided by an even more diverse amalgam of fiber-optic and hybrid-fiber-coax system, plus DSL, cable modems, competitive local exchange carriers (CLECs), gigabit ethernet, and even power line communications systems. These various wire-based systems are providing broadband services to enterprise networks, small office and home office, and mass consumer services.

At no time in history have telecommunications and networking standards been more important. It is only the various telecommunications and IP networking standard bodies which set the rules and standards to allow high-quality communications and rapid interconnection of different media to occur.

D. New Satellite Architecture and User Terminals

The technical logic behind the Negroponte Flip was, of course, quite valid even though market demand drove services in another direction. The shortage of new spectrum and the difficulty of finding it, through new and more intensive reuse techniques or new frequency allocations, are getting ever more exacting. Broadband wireless market demand has been satisfied so far only by an extremely clever combination of digital compression techniques, demanding new frequency reuse techniques, and difficult reallocations of frequencies, plus some use of new spectrum in the microwave and millimeter wave zones. For wireless systems and particularly satellite systems to remain responsive to the market in the future, there is a clear need for some new technical breakthroughs.

In this book, some ideas concerning totally new types of satellite design and system architecture for 21st century space communications systems will be explored. Major new breakthroughs appear necessary if satellite systems are to be able to satisfy the market demand for broadband services and do so at prices that consumers and enterprises will be willing to pay.

Systems that are tens to hundreds of times more efficient than those of today and with much higher throughput capabilities will be needed to respond to the market demands predicted a decade or so from now.

E. B2B vs Consumer Demand

In the early years, the world of satellite services was driven by the needs of large telecommunications carriers to broadcast television over wide areas and to provide transoceanic telephone and data services for both consumers and business. Today, satellite services are no longer simply the adjuncts of telecommunications monopolies but rather mechanisms to provide B2B connectivity among enterprise networks to meet internal and external communications needs. The satellite industry is also increasingly able to support direct access services to consumers as well as small office and home office users. A new type of satellite architecture has been introduced by Hughes Directway, SES ASTRA hybrid Ku-and Ka-band services, and Gilat Spacenet and Starband. The generic form of this new architecture is illustrated in Fig. 1.6.

The question thus arises: "Is the future of satellite networks dependent on business enterprise requirements or largely on consumer oriented communications and entertainment services?" The answer is yes!

Satellites, to be commercially successful, must be able to support both types of customers on a flexible basis. Very high-powered satellites, low-cost microterminals, and new digital (especially IP-based) services are becoming available just at the right time to allow satellites to increasingly provide a panorama of services to satisfy the needs of both businesses and consumers. New products such as the Gilat Spacenet terminal, which can support the reception of direct broadcast television entertainment from two Echostar satellites and also support high-speed interactive broadband services via other adjacent satellites, are indicative of the innovative market integration expected in the future. This integration will not always be easy. Business systems offer a higher value (and thus higher-priced service) that includes such elements as high-level encryption, higher quality of service (or lower bit error rates), increased system availability, networking interconnectivity, and other specially encoded, dynamic, and on-demand systems such as video conferencing on an as-needed basis. The key in all this is very much in the design of ground systems.

There is currently great emphasis on trying to design flexible ground systems to meet a variety of business and consumer demands. In time, however, there may well be "smart" microterminals able to integrate various services. For consumer-based services, these microterminals might even shrink to the size of wearable systems.

F. Regionalism and New Patterns of Globalization

The pattern of satellite communications integration and mergers is now firmly established. During the last five years new domestic and regional entities have emerged, but the dominant trend has been a move toward global integration or at

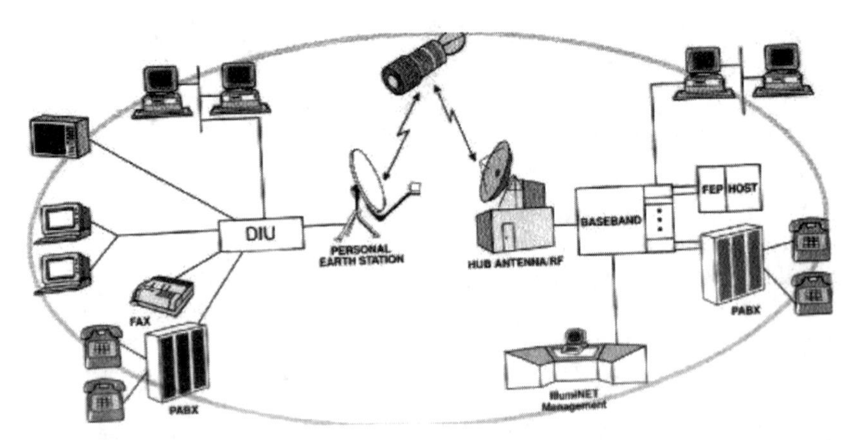

Fig. 1.6 New type of desktop satellite service that can serve B2B modern enterprise networking needs.[4]

least strategic partnerships. One of the more dramatic occurrences is that of SES GLOBAL, which now owns the former GE Americom system, the former Columbia Satellite System, the Astra Satellite System, and has partial ownership of the Asiasat Satellite System, the South American Star One system, Nahuelsat and the AAP-1 satellite, and the Scandinavian NSAB system. In all, the entire SES GLOBAL fleet includes some 43 satellites spread all over the world.

Further changes have occurred recently, and even more are on the horizon. At the time of publication, Intelsat had acquired the investment in its system that was once held by Comsat/Lockheed Martin.[4] Lockheed Martin is selling off its other satellite holdings to other satellite systems, with Telenor acquiring the former Comsat holding in Inmarsat. It appears extremely likely that other mergers and acquisitions will occur. Echostar may acquire the Panamsat and Hughes Galaxy, although its acquisition of the DirecTV satellite systems has now been blocked by the U.S. regulatory authorities. Eutelsat and Hispasat might merge further together or otherwise be acquired, and perhaps New Skies might merge with another entity. The trend toward merger and institutional consolidation thus appears likely to continue throughout the satellite service and manufacturing industries for some time to come.

Separate from changes of institutional arrangements, privatization, and mergers, other types of change at the global level should also be particularly noted. Japan has become the first developed country to reach zero population growth, and Europe appears likely to be the next region to follow suit. In one part of the world, the so-called developed economies, one sees extremely rapidly expanding demand for information and IT services, particularly Internet-related IP services, but very slow population growth. In other parts of the world, the so-called newly industrializing countries, there is even more rapid IT and telecommunications-related growth and, in many cases, continued population growth. Finally, in the least developed countries, there is as yet modest demand for new IT and telecommunications services but rapid population growth and rapid expansion of Internet services.

The idea that one type of telecommunications system or one type of satellite system can respond to the needs of all the various parts of the world would be incorrect. Satellite system operators at the national, regional, and global levels must be flexible and able to react to shifting patterns of demand. This need for flexibility, unlike the super speed trend that suggests bigger and bigger satellites with higher economies of scale, actually seems to suggest the opposite idea. Flexibility can in part be achieved by developing different types of user terminals that have broader access to different types of satellites. Certainly the idea of having universal access user terminals would seem to offer businesses and residential users advantages, even though satellite owners and operators might resist such a concept.

V. Overview of this Publication

This book is designed to provide an introduction to key technical concepts, an understanding of the evolving satellite market on a global scale, an explanation of the various satellite services (including fixed, mobile, broadcasting, broadband Internet, and other offerings), and a presentation of current trends to develop improved technologies and services in each field. Finally, it seeks to explore the longer-term future of satellite communications and the strategies that can be pursued on that time scale to make satellites more cost effective and competitive. To improve the educational value of the book, each chapter includes a series of questions whose answers will help the reader better understand the content of each section.

Chapter 2 provides a foundation with regard to satellite communications technology. It discusses the basics of orbital mechanics, satellite communications, modulation and multiplexing concepts, spacecraft subsystems, launch operations, satellite frequencies, and different types of satellite services that can be provided in existing and future spectrum band allocations.

It is assumed that most readers have some background knowledge of the field, but Chapter 2 provides a foundation from which to read the chapters that follow. It is hoped that the interactive sections at the end of each chapter will provide a basis for deriving a deeper understanding and a better ability to analyze the content provided with regard to fixed, mobile, broadcasting, broadband Internet, and longer-term future satellite systems.

References

[1]*Merrill Lynch Report on Satellite Communications*, Merill Lynch, New York, 2001.

[2]Dataquest, Gartner Group, "Market Survey on Satellite Communications," Sunnyvale, CA, 2001.

[3]International Engineering Consortium, *2001–2002 Annual Review of Communications*, Chicago, 2002.

[4]*The New Satellite Industry: Revenue Opportunities and Strategies for Success*, International Engineering Consortium, Chicago, 2002, pp. 16–18, 25–31.

[5]Staples, G., *Telegeography 2002*, Telegeotraphy Inc., Washington, DC, 2002.

Suggested Reading

Alper, J., and Pelton, J. N. (eds), *The INTELSAT Global Satellite Network*, Progress in Astronautics and Aeronautics, AIAA, New York, 1986.

Aversa, J., "ECC Okays Satellites Providing Two-Way Data Services," *Daily Camera*, Oct. 1994.

Brandon, W. T., "Market Elasticity of Satellite Communications Terminals," *Journal of Space Communications*, Vol. 10, No. 4, 1992, pp. 279–284.

Branscomb, A. (ed.), *Toward a Law of Global Communications Networks*, Longman, New York, 1986.

Bross, D., "Boeing Targets Government for In-Flight Broadband," *Satellite News*, PBI Media, Potomac, MD, 4 Feb. 2002.

Bross, D., "The Year 2002: What's Next for DBS," *Satellite News*, PBI Media, Potomac, MD, 28 Jan. 2002.

Brown, S., "How the Europeans Respond to Mobile Communications," *Communications News*, May 1994, pp. 28–30.

Bull, S., "The Status of the Interactive VSAT Market," *Via Satellite*, Dec. 2001, pp. 34–38.

Bulloch, C., "A Cautiously Optimistic Look at European VSATs," *Via Satellite*, Dec. 1994, pp. 40–43.

Center for Telecommunications Management, "*The Telecom Outlook Report*," International Engineering Consortium, Chicago, 2001.

Deloitte and Touche, "1998, 1999, and 2000 Wireless Communications Industry Survey," Deloitte and Touche, Atlanta, 1993.

Dykewicz, P., "GM Eyes Robust XM Growth From OEM Sales," *Satellite News*, PBI Media, Potomac, MD, 28 Jan. 2002, p. 2.

Dykewicz, P., "Sirius Moves to Escape from XM's Shadow," *Satellite News*, PBI Media, Potomac, MD, 4 Feb. 2002, pp. 1, 8.

Dykewicz, P., "Teledesic Slashes Broadband Plans," *Satellite News*, PBI Media, Potomac, MD, 11 Feb. 2002, pp. 1, 8.

Dykewicz, P., "World Space Weighs Delaying Ameristar Launch," *Satellite News*, PBI Media, Potomac, MD, 28 Jan. 2002, pp. 1, 3.

Frieden, R., "Wireline vs. Wireless: Can Network Parity Be Reached?" *Satellite Communications*, July 1994, pp. 20–23.

Glaser, P., "The Practical Uses of High-Altitude Long-Endurance Platforms," International Astronautical Federation 44th Congress, Graz, Austria, Oct. 1993.

Gordon, G. D., and Morgan, W., *Principles of Communications Satellites*, Wiley, New York, 1993.

Horwitz, C., "The Rise of Global VSAT Networks," *Satellite Communications*, April 1994, pp. 31–36.

ITU International Table of Frequency Allocations, Part 2 of FCC Rules and Regulations, Government Printing Office, Washington, DC, 1999.

Iida, T. (ed.), *Satellite Communications-System and its Design Technology*, Ohmsha Publishers, Tokyo, and IOS Press, Amsterdam, 2000.

Iida, T., "Designing Low-Cost Satellite Communications Systems for Remote Tele-Science," Lecture in ISU 1994 Summer Program.

Mason, C., "Will the U.S. Remain Competitive in the Wireless Future?" *Telephony*, 12 July 1998.

Meagher, C., *Satellite Regulatory Compendium*, Phillips Publishing, Potomac, MD, 1997.

NASA/NSF, *Global Communications Satellite Technology and Systems*, International Technology Research Institute, Baltimore, MD, Dec. 1998.

NASA/NSF, "Panel Report on Satellite Communications Systems and Technology," Vols. 1 and 2, International Technology Research Institute, Baltimore, MD, July 1993.

National Science and Technology Council, "Networked Computing for the 21st Century," Subcommittee on Computing, Information and Communications R&D, Washington, DC, Sept. 1998.

Newton, H., *Newton's Telecommunications Dictionary*, Telecommunications Library, New York, 1999.

Ohmori, S., Wakana, H., and Kawase, S. (eds.), *Mobile Satellite Communications*, Artech House, Boston, MA, 1998.

Paschall, L. M., "Security Aspects of Satellite and Cable Systems," *Journal of Space Communications*, Vol. 6, No. 4, Amsterdam, 1988, pp. 269–276.

Payne, S. (ed.), *International Satellite Directory*, Design Publishers, Sonoma, CA, 2001.

Pelton, J. N., "Alternative Visions of Satellite Communication Economics in the 2010–2020 Time Period," IAF World Space Congress, Houston, TX, Oct. 2002.

Pelton, J. N., "The New Satellite Industry: Revenue Opportunities and Strategies for Success," International Engineering Consortium, Chicago, 2002.
Pelton, J. N., *Wireless and Satellite Telecommunications*, Prentice-Hall, Upper Saddle River, NJ, 1995.
Pelton, J. N., "Advanced Concepts in Satellite Communications," *Proceedings of MILCOM 2001*, Fairfax, VA, 29 Oct. 2001.
Pelton, J. N., "Five Ways Nicholas Negroponte Is Wrong About the Future of Telecommunications," *Telecommunications*, Vol. 11, No. 4, Apr. 1993.
Pelton, J. N., "Overview of Satellite Communications: Global Trends in Deregulation and Competition," International Astronautical Federation, 45th Congress, Jerusalem, Oct. 1994.
Rogers, J., "Sirus Moves to Escape from XM's Shadow," *Space News*, Feb. 2002.
Schnaars, S., *Megamistakes*, MacMillian, New York, 1986.
"Export Control Issue Again in Spotlight," *Satellite News*, 11 Feb. 2002, p. 7.
"Reformed ITU Filing Procedures—Brokers for Orbital Space-Boon or Bane?" *Via Satellite*, May 1994.
"The Communicopia Study: C-5 Convergence," Goldman-Sachs, New York, 1992.
"The DVR Market: The Good, The Bad and the Ugly," *Satellite News*, 11 Feb. 2002, pp. 4–5.
"The Direct TV DBS System," *Via Satellite*, 1993.
"The Information Wave: Digital Compression is Expanding the Definition of Modern Telecommunications," *Uplink*, Spring 1994, pp. 4–7.

Questions for Discussion

1) The rate at which information is doubling worldwide is estimated to be every three years, and the time required for the doubling rate continues to shrink. This means that there may be at least 16 times more information to process, store, transmit, and receive by 2015. Further, the demand for mobile broadband access to information databases and entertainment is soaring. The potential impact of this information explosion is enormous.

   What special impact on the satellite industry could the continuing explosion of information worldwide provide? Could satellites play a special role in the creation of a "worldwide mind," a global information exchange, or on an "always-on" global electronic university?

2) The information carrying capability of lasers may well rise to the petabits/s capability in coming decades and, in terms of trunk or heavy-duty connections of cities, fiber-optic cables seem destined to play a key role for the future. However, in the areas of mobile broadband communications, and in broadcasting and multicasting applications in health, education, and government services, satellites seem to offer great value and flexibility especially in creating instantaneous connectivity on a global basis.

   How will satellites make their most important contributions to world society 20 years from now and will this be in urban, suburban, remote, or rural areas?

3) It is suggested in this chapter that technical standards which allow various types of telecommunications media (that is, fiber, coax, terrestrial wireless, and satellites) to interconnect seamlessly are key. It is further suggested that the effective linkage between the networking IP world and the telecom world is key to the future of satellite communications.

   Do you agree? What problems, such as satellite latency, do you see as barriers to effective global and universal technical standards? Does the gap between the technical standards setting process by the International Telecommunication Union, ANSI, ETSI, and CITEL, on the telecom side and the Internet Engineering Task Force on the IP networking side seem to represent a key problem or issue in this regard? How might these problems be better addressed?

4) The World Trade Organization, as well as various regional trade and economic unions such as the European Union, NAFTA, and APEC, have moved to open free trade in services including telecommunications, networking, and satellites. Under the General Agreement on Trade in Telecommunications Services, more and more countries have agreed to open their borders to telecommunications competition and have removed barriers to competition.

   Do you agree that free trade is key to the future growth of global and regional satellite systems? Why would this be particularly true for satellites as opposed to other media such as fiber-optic cables or terrestrial wireless services? Are you aware of the many regulatory issues that satellites must currently address to provide international telecommunications services such as allocation of frequencies at the global and regional level, the licensing of frequencies at the national level, the granting of national landing rights, and the different procedures that apply to providing either basic backbone services and/or value-added services to end users?

5) Satellite communications technology and services have rapidly advanced in the last few decades but progress has been uneven as an industry. There have been parts that have experienced major growth and economic advances while others have experienced major setbacks and even some bankruptcies.

   Which parts of the satellite industry [i.e., fixed satellite services (FSS), mobile satellite services (MSS), broadcast satellite services (BSS), messaging, space navigation, GPS, and broadband Internet services] do you perceive as succeeding and which of these have been much less successful? What do you believe are the keys to success for the FSS and BSS networks? Why do you believe that MSS and early broadband satellite systems have been much less successful? How can better technology, better orbits, better business plans, or better regulatory and trade policies help?

6) There has been a wave of consolidations, mergers, and acquisitions going on in the field of satellite communications. These moves toward consolidation seem to be driven by the urge to merge costly facilities and research programs, seek economies of scale, increase marketing power, and seek other cost-efficiency advantages. Several satellite manufacturers, satellite operators, and research organizations have merged or been acquired, and more such changes are likely to occur.

   Do you think this process strengthens or weakens research and increases or decreases research spending? Will this process ultimately increase or decrease the cost of satellite services? Do you feel this process is necessary for satellite operators to become cost competitive? Do you think the issue of GEO vs MEO and LEO orbit satellite systems is related to this trend; if so, how? Do you think the possible consolidation of fixed, broadband Internet, and broadcast entertainment services might occur on future digitally based systems; if so, do you think this will affect the consolidation and merger process?

Chapter 2

Principles of Satellite Communications Systems

Edward Ashford*
SES GLOBAL, Betzdorf, Luxembourg

I. Introduction

THIS book is not intended to be a textbook on either communications engineering or satellite communications. However, to understand the reasons why many of the technological developments mentioned in this book are or will become required, and to obtain an appreciation of the effects that these will have, it is necessary to have a certain level of understanding of the principles underlying any communications system, and of satellite communications in particular. Furthermore, many of those not working in the fields of space or communications may appreciate an introduction to the terminology (some might say jargon) used in the field. This chapter, therefore, is intended to introduce readers, who may not feel that they already have the necessary technical background, to certain basic principles and terminology that it is hoped will make the material in the following chapters more understandable. Many readers will already have a knowledge of the field of satellite communications, either obtained academically or through working in it or in a related field. In such cases, this chapter may appear too basic; these readers may wish to skip this chapter and go on to the next.

Finally, the authors of this book have intentionally tried to limit the use of mathematical equations to make it, as much as possible, understandable to non-technical readers. Nevertheless, many of the concepts underlying satellite and communications technology require the use of at least a limited number of equations to explain the relationships between the various parameters involved. For these, equations will be used, but the surrounding text will attempt to explain these in relatively simple terms.

II. Why Communications Satellites?

Until the advent of the first communications satellites in the first half of the 1960s, communications were normally handled by either direct wire connections

*Vice President for Technology Development.

or by terrestrial radio frequency (RF) broadcasting. Since that time, however, many hundreds of satellites have been developed and launched to provide communications services. Not only is the number of satellites impressive, but also impressive is the continuous evolution of satellite capabilities over a relatively short period. For example, the first commercial geostationary Earth orbit (GEO) communications satellite, Syncom-3, was able to carry only one television channel when it was put into service in 1964. A single GEO satellite launched today might be able to carry 400 or more channels.

Why have communications satellites become so much a part of our everyday life? There are many reasons, but perhaps the most significant is that they can provide communications services to very large geographic areas. Satellites, because they are high above the surface of the Earth, can receive and transmit signals from huge areas, and the higher they are, the greater these areas become. Furthermore, it is often far quicker and less expensive to bring broadcasting and communications to areas lacking them using satellites than it would be to provide the same services solely by terrestrial means.

III. Satellite Orbits

Early in the 17th century, the German astronomer Johannes Kepler formulated and published three empirical laws that described how (but not why) planets revolved around the sun. Kepler's laws can be paraphrased as:[1]

1) The orbit of a planet around the sun has the shape of an ellipse, with the sun lying at one of its two foci.
2) A line connecting a planet with the sun (its instantaneous radius vector) "sweeps out" equal areas in any fixed increment of time as the planet revolves around the sun.
3) The square of the time taken by a planet to revolve once around the sun (its period) is proportional to the cube of its average distance from the sun (the semi-major axis of the ellipse).

Figure 2.1 illustrates the first two of these laws, showing the elliptical shape of a planet's orbit around the sun and an area (cross-hatched) being swept out by its radius vector in an increment of time, Δt. The second law, in particular, is a reflection of the fact that the velocity of a planet in orbit varies with time, and is greater when the planet is nearer the sun (greatest at perigee) than when it is farthest away (slowest at apogee). The figure also shows the meaning of semi-major axis, denoted by a, and introduces another parameter, called the eccentricity of the orbit and denoted by e, which is a measure of just how elliptical an elliptical orbit is.

In the latter part of the 17th century, Issac Newton published his famous Three Laws of Motion, and his Law of Universal Gravitation. These laws can be used to derive each of Kepler's empirical laws, thereby explaining the "why" that Kepler's laws did not address. Using Newton's laws, one can derive an equation for a planetary orbit:[2]

$$r = \left\{\frac{1}{1 + e\cos\theta}\right\}\left\{\frac{\mathrm{h}^2}{\mu}\right\} \tag{2.1}$$

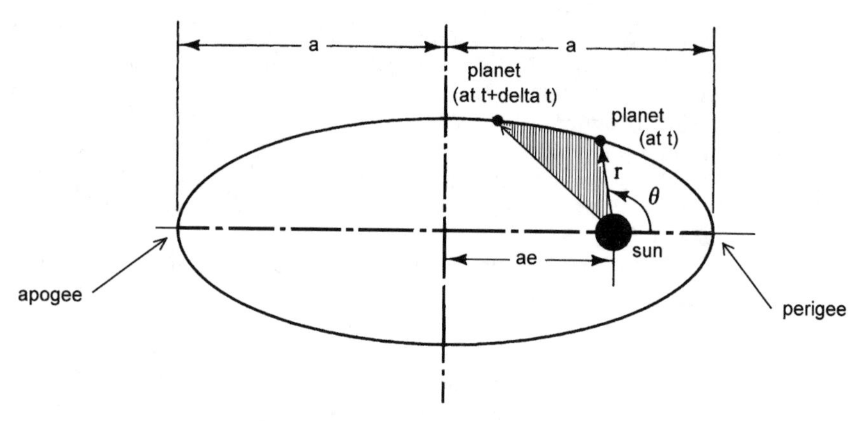

Fig. 2.1 Kepler's planetary orbits.

where h is a constant for a specific orbit, related to its angular momentum, μ is the product of the universal gravitational constant, $G\ (= 6.673 \times 10^{-20} \text{km}^3/\text{kg}\cdot\text{s}^2)$ times the mass of the central body, M, and e is the eccentricity parameter mentioned above. For Earth, $\mu = 398{,}601.2 \text{km}^3/\text{s}^2$.[2]

In the previous text, I have used the term central body instead of the sun, because Newton generalized Kepler's laws to apply to any case where a relatively small body (such as a communications satellite) orbits around a much more massive body (such as the Earth). Thus Eq. (2.1) also applies to planets, communications satellites, planetary probes, or the Apollo lunar missions of the 1960s. It does not apply, however, at least strictly, to the orbit of the moon around the Earth, because the mass of the moon is not relatively small enough with respect to that of the Earth.

Equation (2.1) is more general than Kepler's laws, which only apply to a body orbiting around a large central body. Depending on the value of the eccentricity parameter, the equation can apply to several types of orbit. When e is equal to zero, it reduces to defining r as a constant, which is the equation for a circular orbit (a special case of Kepler's elliptical orbits, when the two foci of the ellipse coincide). If e becomes equal to 1.0, the equation indicates that the angle θ cannot equal 180 deg, because that would imply a value of zero in the denominator, which is forbidden in mathematics. The equation then becomes one of a parabola rather than an ellipse, which could apply, for example, when a planetary probe satellite approaches a planet with a particular velocity and direction, is deflected under the influence of its gravitational pull, swings around it, and then continues onward into outer space. If e is greater than 1.0, there are two excluded values for θ, and the equation becomes one for a hyperbola. This could also apply for the planetary probe previously mentioned, when the approach velocity to the planet is greater than that which would produce a parabolic orbit.

In both of the latter two cases, it is debatable whether one should refer to the equation as being that of an orbit or of a trajectory. In further discussions of communications satellites, however, only circular and elliptical orbits will be considered, so this point does not concern us.

Satellites can provide useful communications services from a variety of different types of orbit. Figure 2.2 shows a typical designation of some important orbits, as being low Earth orbits (LEO), medium Earth orbits (MEO), intermediate circular orbits (ICO), highly inclined elliptical orbit (HEO), and the GEO. Before looking further into the details of how communications satellites can use these types of orbits in later chapters, it would be useful to first understand some of the differences between various orbits that would influence a satellite designer to choose one type over another.

In Fig. 2.2, only one of the orbits, GEO, has a specific radius from the center of the Earth associated with it, that being 42,164 km. For the MEO and LEO cases, there are, in general, an infinite number of possible orbits, each having different altitudes, eccentricities, and inclinations with respect to the Earth's equator. There is, however, only one strictly geostationary orbital altitude. Arthur C. Clarke, sometime referred to as the father of communications satellites, pointed out the benefits of satellites in high orbits in an article.[3] He noted that a satellite in a circular equatorial orbit at an altitude of 35,786 km (altitude above the surface of the Earth, and hence at a radius of 42,164 km), would take just as long to complete one full orbit as it takes the Earth to rotate 360 deg. A satellite in such an orbit would therefore seem, to an observer on Earth, to remain fixed in space. More than one-third of the globe would be visible from that altitude, and therefore anyone in that area of visibility could transmit to or receive from the satellite. Three such satellites, equally spaced in their orbits around the equator, would be able to cover almost all the populated area of the Earth. Figure 2.3, which is a view as seen from a vantage point high above the North Pole, illustrates this principle. In Fig. 2.3, the shaded portions indicate where two-satellite coverage is available. The small circle denotes the region above approximately 81° north latitude (or below 81° south latitude) within which a satellite in GEO is below the local horizon and is therefore not visible.

What is the magic that causes a satellite in a circular equatorial orbit at an altitude of 35,768 km to appear to be fixed in the sky? From Kepler's Third law and

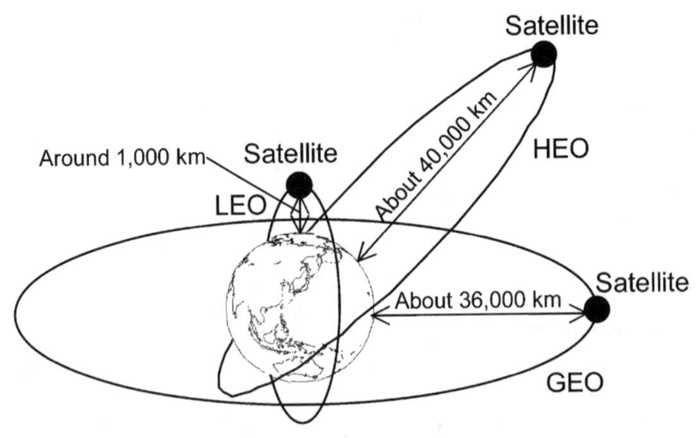

Fig. 2.2 Different types of communications satellite orbits.

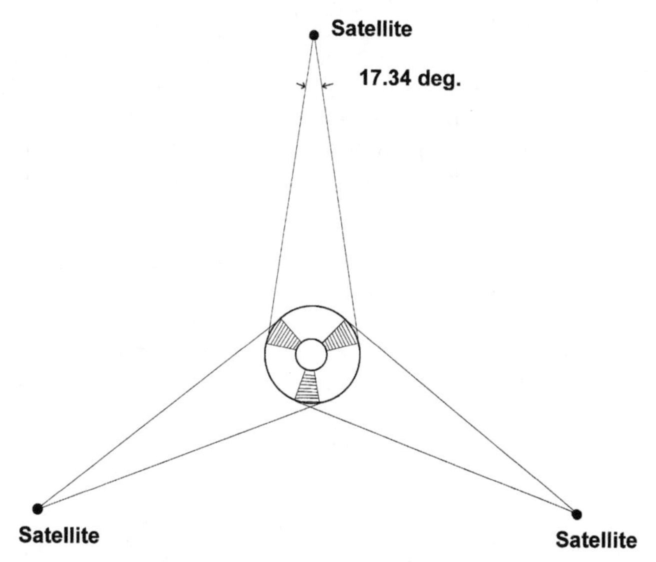

Fig. 2.3 Three-satellite coverage from the Clarke GEO orbit.

Newton's Law of Gravitation, one can derive an equation for the period of any circular orbit in terms of its radius:[4]

$$T = 2\pi\left\{\frac{r^3}{\mu}\right\}^{1/2} \tag{2.2}$$

For a satellite to appear stationary to an observer on the ground, it must be in a circular orbit directly over the equator with a period of rotation exactly that of the Earth. Because the Earth is moving about the sun as it rotates about its North–South polar axis, the satellite period needed to seem stationary is not exactly 24 h (the average period of the Earth's rotation from sunrise to sunrise), but is somewhat less (23 h, 56 min, and 4.09 s, to be more precise).[2] If this value is used in Eq. (2.2), the calculated circular orbit radius (measured from the center of the Earth) will be 42,164.16 km. Because the average radius of the Earth at the equator is some 6396 km, the altitude of the GEO orbit is approximately 35,768 km.

Equation (2.2) applies strictly to the case where a very small body orbits around a much larger body with a spherically symmetric gravitational field. In the real world, this is not quite the case. The Earth is not a perfect uniform sphere, but is slightly pear-shaped. Furthermore, the density of the Earth's crust and mantle varies slightly from place to place. For both these reasons, the Earth's gravitational field does not exert a constant attraction toward the center of the Earth on a satellite in a circular orbit about it. Nonuniformities in the gravitational field can influence the orbit of a satellite in almost any orbit. In the case of the GEO orbit, the major effect is that a satellite that is initially stationary as seen from the ground will gradually drift away from that position, in an Easterly or Westerly direction. The direction and magnitude of this drift depends on the satellite's longitude over the equator, and points of stable equilibrium exist at longitudes of approximately 105° West and 75° East,[5] but whatever it is, the effect is relatively weak, and satellites can be brought back to their desired longitude relatively easily by firing small thrusters (tiny liquid fuel rocket engines) periodically. The maximum annual velocity increment that thrusters must impart to a GEO satellite to maintain its longitudinal position over the equator, a procedure known as East–West station keeping, is approximately 2 m/s (Ref. 6).

Of more importance is the fact that the Earth's is not the only gravitational field that affects the orbit of one of its satellites. The gravitational attraction of both the sun and moon also act on the satellite. Their principal effect, in the case of a GEO satellite, is to cause its inclination to gradually increase. If not corrected, this effect will cause the inclination of a GEO satellite to increase between approximately 0.75 and 0.94 deg per year, depending on the year in question. If this effect is not corrected, the inclination will gradually increase, up to a maximum value of approximately 14.7 deg in approximately 27 years. Thereafter, the inclination would begin to decrease again, oscillating between 14.7 and 0 deg with a period of approximately 54 years.[5] As in the case of East–West station keeping, thrusters are used to maintain a GEO satellite's inclination near 0. This North–South station keeping is much more costly in terms of thruster propellant, because a velocity increment of as much as approximately 51 m/s is required during the years when the buildup of inclination is at its peak.

Because satellites at higher orbits can see greater percentages of the total Earth, as the altitude above Earth of satellite orbits increases, fewer satellites are required to provide continuous uninterrupted 24-h coverage over the entire portion of the globe visible from that orbit. This is illustrated in Fig. 2.4.

Not all altitudes are good choices at which to locate a satellite's orbit. In addition to light, the sun emits large quantities of electrons and protons at high speeds that, together, constitute what is known as the solar wind. When these approach the Earth they are deflected by the Earth's magnetic field in such a way that they tend to become concentrated in vaguely doughnut-shaped layers surrounding the equator. These layers are called the Van Allen belts, after the scientist who discovered and

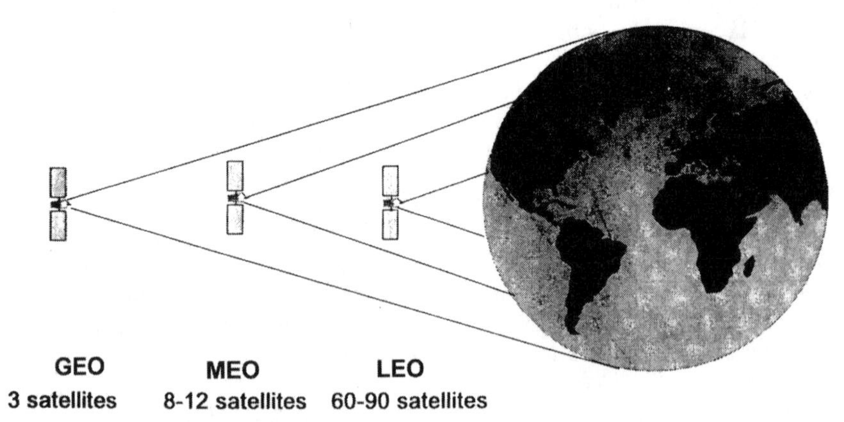

Fig. 2.4 Earth coverage vs satellite orbital altitude.

analyzed them. High concentrations of protons become concentrated in a region beginning at approximately 1000 km and extending out to between 9000 and 10,000 km, with peak intensities around 3700 km. High concentrations of electrons begin at approximately the same altitude, but extend out much further, to 30,000 km or more, peaking at approximately 21,000 km (Ref. 6).

The solar cells used to provide power to satellites are subject to damage from exposure to such proton and electron irradiation. Even the integrated circuits used in the various satellite subsystems can be adversely affected by exposure to these high particles. Of the two, protons are typically the most damaging, but both types of exposure can cause damage. Thus, satellite designers prefer to avoid regions of high radiation and, if possible, choose orbits that exclude these. Thus LEO satellite orbits are usually selected to be between 300 and 1500 km altitude (the lower value in the range chosen not because of radiation but to avoid atmospheric drag having too great an influence on the orbit parameters), while MEO/ICO satellites usually have orbits above 10,000 or 15,000 km.

HEO satellites may have their perigee near the altitude of a LEO orbit and their apogees at MEO (or even GEO) altitude. Because the intensity of the Van Allen belts falls off the further one gets from the equatorial plane, the fact that the orbits have a high inclination (with respect to that plane) helps to reduce their radiation exposure. Also, because for such an orbit the velocity while in the lower (proton) belt is relatively high (being near perigee) will reduce the time of exposure and hence the overall integrated radiation dose seen by the satellite. Even with these two mitigating factors, however, HEO satellites are generally considered to have a shorter lifespan, because of radiation damage, than do satellites in LEO/MEO/GEO orbits.

IV. Concepts Involved in Satellite Communications

The basic concept underlying any communications system is that of the communications link. While each part of it can be highly complex, in its simplest form a link consists of only three conceptual elements: a transmitter, a transmission medium, and a receiver. The link is designed to convey information from one place to another; baseband signals carry this information.

A link may convey information only in one direction (called a simplex link), which would be typical of a broadcast link for example, or in two directions (called a duplex link), with each end of the link having both a receiver and transmitter. A telephone link, for example, in its simplest form, is a duplex link consisting of two handsets, each having a microphone (the transmitter) and an earphone (the receiver), interconnected by copper wires (the transmission medium) to carry the electrical signals used by the microphones and earphones from one to the other.

Where satellite communications are concerned, two links are usually considered to be involved. Figure 2.5 shows the example of a television broadcast satellite, showing a link from a transmitting terminal on Earth to a receiver on the satellite (the uplink), and a second link from a transmitter on the satellite to a TV receive-only (TVRO) terminal on the ground (the downlink). On board the satellite, two types of actions may occur. In one case, one or more modulated carriers transmitted from the ground may be received, shifted in frequency (to

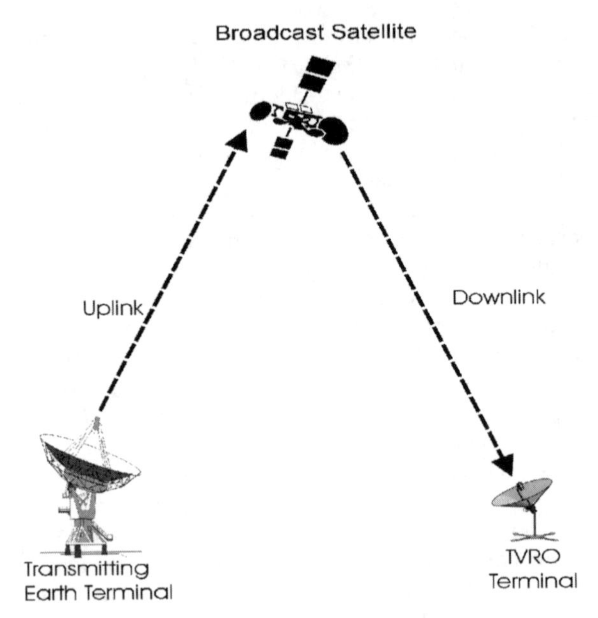

Fig. 2.5 Links for satellite transmissions.

avoid interfering with the received signals), amplified, and retransmitted to receiving terminals on the ground. This is an example of what is referred to as a bent-pipe transmission, using the analogy of a satellite acting as a water pipe where the water (content) going in is the same as that coming out, having merely been changed in the direction it is moving. The satellite payload is referred to as a bent-pipe or transparent payload, or more commonly as a repeater.

Alternatively, the modulated carrier received at the satellite may be demodulated on board the satellite to recover the information content. Certain actions may then be taken with the content before it is retransmitted back to the ground, which can enhance it in some way or direct it toward specific locations on Earth. This latter case is an example of what is known as onboard processing (OBP). What is done on board the satellite, and the type of equipment used to do it, can depend greatly on whether the satellite is to be used for analog or digital transmissions.

A. Modulation

The information conveyed via the transmission medium (the signal or content) can be either analog or digital in format. In analog transmissions, signals vary continuously either in amplitude, frequency, or phase, or in any combination of these. In digital transmissions, signals vary discretely, jumping between dissimilar values to indicate binary “ones” and “zeros.”

To convey the signal, one usually starts with an alternating waveform, of relatively high frequency and having a fixed amplitude and frequency, called a carrier. A signal carrying the desired information (the content), generally at a much lower frequency (the baseband frequency) than the carrier, is used to modify one or more of the characteristics of the carrier (amplitude, frequency, or phase). This process is referred to as modulation. For digital transmissions, some form of coding is often added to the signal to allow a receiver to detect and even correct errors in the received signal caused by noise or other transmission impairments. In either event, the resulting modulated carrier is amplified and then passes over the transmission medium to the receiver, where, if it is the end of the link in question, demodulation takes place, separating out the carrier and the information signal again.

If an analog signal is used to modulate a carrier, the modulated signal can, within certain limits, have any value of amplitude, frequency, or phase angle. If it is a digital modulating signal, the same types of modulation can be used, but rather than there being an infinite continuum of amplitude, frequency, or phase in the modulated carrier, only a limited number of discrete amplitudes, frequencies, or phases will be present. Figure 2.6 (part a) shows, for example, the case where a sinusoidal carrier is amplitude modulated (AM) by an analog signal. Part b of Fig. 2.6 shows the case of amplitude modulation when the signal is digital instead of analog. This is one type of example of amplitude shift keying (ASK), whose best-known example is perhaps the sending of Morse code over a radio link.

Figure 2.7 shows the case of a carrier's frequency being modulated by a digital signal. Frequency modulation with digital signals is also known as frequency shift keying (FSK). In this case, there are two different carrier frequencies, and the modulator shifts between these two depending on whether a digital "zero" or a digital "one" is to be sent. Finally, Fig. 2.8 displays how a digital signal can modulate the phase of a sinusoidal carrier. Phase modulation is also known as phase shift keying (PSK).[7]

Figure 2.8 shows the case where a digital modulating signal creates one of two discrete states in the modulated carrier, each state having a phase angle

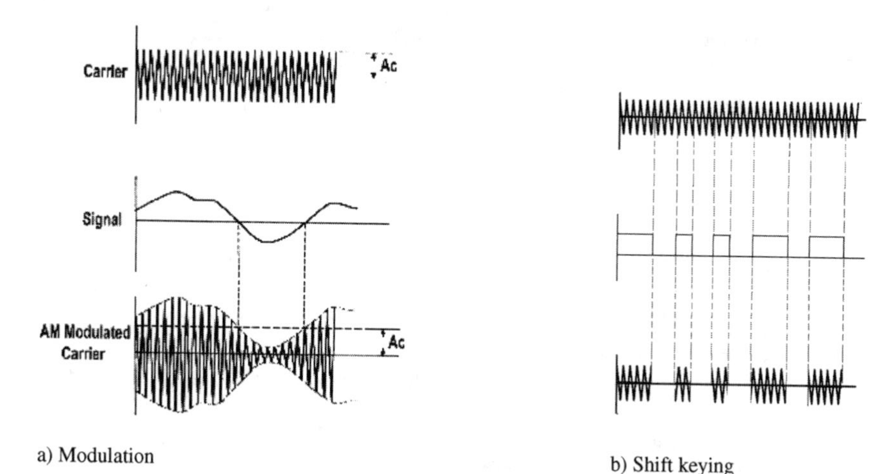

a) Modulation

b) Shift keying

Fig. 2.6 Amplitude modulation.

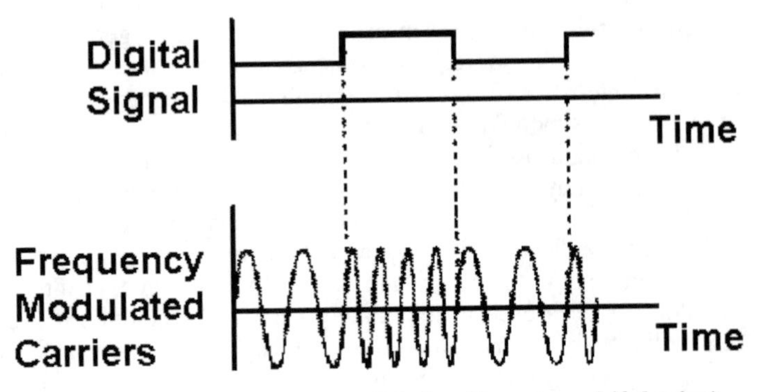

Fig. 2.7 Digital frequency modulation (frequency shift keying).

180 deg apart from the other. This is called biphase shift keying, or BPSK. In fact, what is more common in satellite transmissions is what is called quadrature-phase shift keying (QPSK), where four discrete states are used, spaced at 90-deg phase intervals. Even higher orders of modulation are used in some terrestrial transmissions, although their use in satellites is not yet common. These include 8-PSK, in which the phase shift between symbol sets is 45 deg, 16 QAM, where 16 different states have both different phase shifts and different amplitudes, and even higher orders of each of these.

The term spectrum efficiency, for digital transmissions, refers to how many bits per second (bps) can be transmitted per Hertz of carrier bandwidth. Higher-order modulations give, in general, higher spectrum efficiencies. BPSK has a theoretical efficiency capability of 1 bps/Hz; QPSK has an upper limit of 2 bps/Hz, 8PSK up to 3 bps/Hz, and so on.

One might imagine, therefore, that practically any level of spectrum efficiency could be achieved by simply choosing a high enough order of modulation. Unfortunately, this is not the only consideration. An upper bound on the amount of information, C (in bps) that can be sent through a channel having a bandwidth of B Hz, a signal power level of S and a noise level of N (where S and N are in the same units) is given by what is known as the Shannon–Hartley law:

$$C = B \log_2\left(1 + \frac{S}{N}\right) \tag{2.3}$$

If we maintain a constant output power level on a transmitter, and use progressively higher orders of modulation, the relative power level of each bit of information decreases. Thus, as we attempt to send more information per hertz, the information being sent becomes more susceptible to being corrupted by noise. This process can actually result in being able to send less information through the channel instead of more, as was intended.

As we go to progressively higher orders of modulation, the power level of the signal must be raised significantly to protect against the effects of noise interference. Because power is at a premium in satellites, modulation levels higher

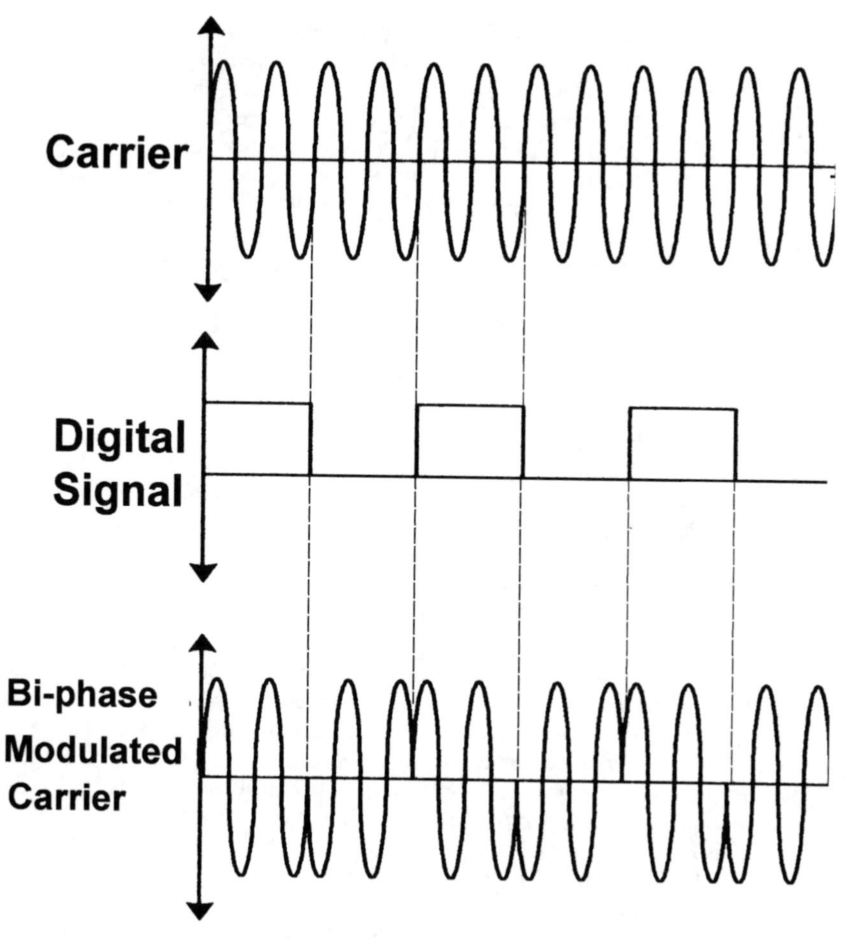

Fig. 2.8 Biphase modulated waveform.

than QPSK are typically used only when the transmission environment can be assured to be relatively low in noise level.

B. Coding

While the topic of coding is a complex one whose details are beyond the scope of this book, it is another important technique used in modern-day digital communications that should at least be understood conceptually by the reader. RF transmissions are prone to errors because of noise or transmission impairments. While digitally modulated signals are less susceptible than analog signals to such noise, if the noise or interference is bad enough even digital transmissions can become difficult or impossible to demodulate without errors. Digital signal transmissions may, however, be coded to allow the receiving terminals to detect errors. A receiver, after an error has been detected, can then send a signal back asking for that part of the transmission to be retransmitted.

Perhaps the simplest example of an error detection code is the use of parity bits. Digital signals, at their simplest, consist basically of a stream of signals in one of two states, referred to as bits (for example, “ones” or “zeros,” on or off, frequency “A” or frequency “B,” or in-phase or 180deg out of phase). If such a stream is divided into groups of, for example, 7 bits each, then in each group there will be either an odd or an even number of “ones” bits. If we add an additional bit to each such group, according to the rule that if there is an odd number of “ones” bits we add a “one,” and if there is an even number of “ones” bits, we add a “zero,” then every group of 8 bits transmitted will have an even number of “ones” bits.

When such a signal arrives at the receiver, each group of 8 bits can be checked to see if it contains an even number of “ones.” If not, the receiver concludes that an error has occurred, and can signal back to the transmitter to request a retransmission of that set of 7 bits. Of course, this example of what is called even parity checking only works if only one of the 7 signal bits in a group has been changed to the other state because of an error. If two bits in the same group have been changed, the check would show that there were still an even number of “ones” received, and the errors would go undetected. There are, however, other types of codes that can be used that allow the detection of higher numbers of errors. They do this by providing additional redundancy to the signal before it is transmitted, so that even multiple errors can be detected.

There are codes, called error correcting codes, which not only can detect but can also even correct errors in transmission. Provided that there are only a limited number of such errors, error correcting codes can automatically correct them at the receiver, thereby avoiding having to ask for a retransmission of the block containing an error. Perhaps the simplest example that illustrates this is the case where we transmit 7, 7-bit words in a block. Each of these words can have a parity bit appended at the end. In addition, each column in the block of 7-bit words can also have parity bits attached, based on whether there are an odd or even number of “ones” in that column. Table 2.1 below shows an example of this.

Now suppose that one bit, say in word 4 in this block of 7 words experiences an error in transmission. Then the parity bit at the end of this word will indicate that

Table 2.1 Simplified dual-parity, error-correcting coding

Column	1	2	3	4	5	6	7	Row parity bits
Word 1	1	0	0	1	1	1	1	1
Word 2	1	0	0	0	1	1	1	0
Word 3	0	0	1	1	1	0	1	0
Word 4	1	1	1	0	0	0	0	1
Word 5	0	0	0	0	1	1	0	0
Word 6	1	1	1	1	1	1	0	0
Word 7	0	0	0	0	1	1	1	1
Column parity bits	0	0	1	1	0	1	0	1

an error has occurred somewhere in that word. Likewise, if the error has occurred in column 4, the parity bit at the end of that column will be in error. The two therefore uniquely indicate both the row and column in which the error has occurred, as is illustrated in Table 2.2. Because there can be only two possible values of the bit in that position, knowing where the error has occurred also allows us to correct it automatically. The bit in row 4 and column 4 must obviously have been a zero rather than a one when it was encoded.

You might ask, "What if a parity bit is an error instead of a data bit?" The example given also allows this to be detected because in that case, with a single error, either a row parity bit will indicate an error has occurred, or a column parity bit will, but not both. In this case, we know that a parity bit was in error, but because we can assume therefore that all the data bits are correct, no error correction will be needed.

Of course, this simple example works only if there is not more than one error. There are, however, types of coding with additional added redundancy that can detect, and correct, several errors in a block. Furthermore, these can work with blocks much longer than the 7-bit by 7-bit block of the simple example. These codes, by the way, are referred to as forward error correcting (FEC) codes because the information needed to correct errors is forwarded to the receiver along with the data. Furthermore, the example is known as a 7/8 FEC code, because for every 7 bits to enter the encoder at the transmitting end, 8 bits exit it and are sent.

The previous example is what is known as a block code. That is, it looks at a block of data at a time, and adds coding bits according to a particular coding algorithm (a rule or set of rules specifying how to encode a block of data). One of the most common block codes is known as the Reed–Solomon code, which, in a typical application, adds 16 parity bits to every string of 188 data bits.[7]

There is another type of coding, called convolutional coding, that looks not only at the data bits in a particular block, but also looks at the bits in the block (or blocks) sent before the one in question, and possibly also at blocks sent afterward. The two-dimensional parity checks such as those described in the example of Tables 2.1 and 2.2 can then be supplemented with a third dimension of checks, in time.

Table 2.2 Detection of a single-bit error

Column	1	2	3	4	5	6	7	Row parity bits
Word 1	1	0	0	1	1	1	1	1
Word 2	1	0	0	0	1	1	1	0
Word 3	0	0	1	1	1	0	1	0
Word 4	1	1	1	**1**	0	0	0	1
Word 5	0	0	0	0	1	1	0	0
Word 6	1	1	1	1	1	1	0	0
Word 7	0	0	0	0	1	1	1	1
Column parity bits	0	0	1	1	0	1	0	

In practice, the two types of coding, block codes and convolutional codes, are each good at detecting and correcting errors caused by different effects. Thus, it is often the case that both types of coding are used in series. The first, or inner coding process, provides a string of bits that are then subjected to a second, or outer coding process. When this is done, the two codes are said to be concatenated.

The addition of coding bits to a signal bit stream is an extra overhead on the digital signal. Because only a certain maximum number of bits can be sent each second via a transmitter having a fixed power level and bandwidth, the addition of coding bits means that fewer signal bits can be sent in the same time. Nevertheless, if the goal is to obtain a received signal with a very high probability that there are no errors, then coding techniques are invaluable in helping to achieve this.

C. Frequency Allocations and Terminology

RF transmissions, of whatever frequency, are merely a relatively narrow part of a continuous spectrum of electromagnetic radiation, as shown in Fig. 2.9 (Ref. 8). Communications satellites are not the only users of the RF spectrum for transmissions. Terrestrial radio and television stations, radars, ham radios, police and emergency vehicle radios, military field communications, airplane communications, and a myriad of other users also need RF spectrum. To avoid all users interfering with one another, the member states of the International Telecommunications Union (ITU), an arm of the United Nations, have agreed on allocations of certain blocks of frequencies for different purposes, for use by different types of users. Some bands of frequencies are assigned for the exclusive use of satellites, while in other frequency bands satellites are not allowed to operate at all. A third category of bands can be used by either satellites or by terrestrial transmitters, provided that the users agree among themselves how to divide up the spectrum in those bands to avoid interfering with one another. Within the bands in which satellites are allowed to operate, there is a further division between the various types of satellite services that exist. Satellites intended to provide fixed satellite services (FSS), that is, to provide links between Earth terminals that are in fixed locations on Earth, have one set of assigned frequencies. Those intended to provide mobile satellite services (or MSS, that is, communications from or to moveable terminals) are assigned another set, and those providing broadcast satellite services (BSS), intended to broadcast to multiple receiving stations, are assigned still another set of frequency bands. Data relay satellites, which use intersatellite links to relay data from one satellite via another to the ground, are assigned yet another set of frequencies. These distinctions between band according to the type of service invisaged are, however, based on their originally intended use when they were set up by the ITU. Particularly with the advent of digital communications, the distinction between them is becoming progressively more blurred, and it is not precluded, for instance, that any of the band could be used to carry broadcasting traffic.

Specific frequency bands have been assigned specific names to avoid having to spell out a long series of numerical digits when referring to them. These names have been assigned during a period while improvements in technology and crowding of the spectrum have pushed transmissions to use higher and higher frequencies. This has caused the names of the progressively higher and higher

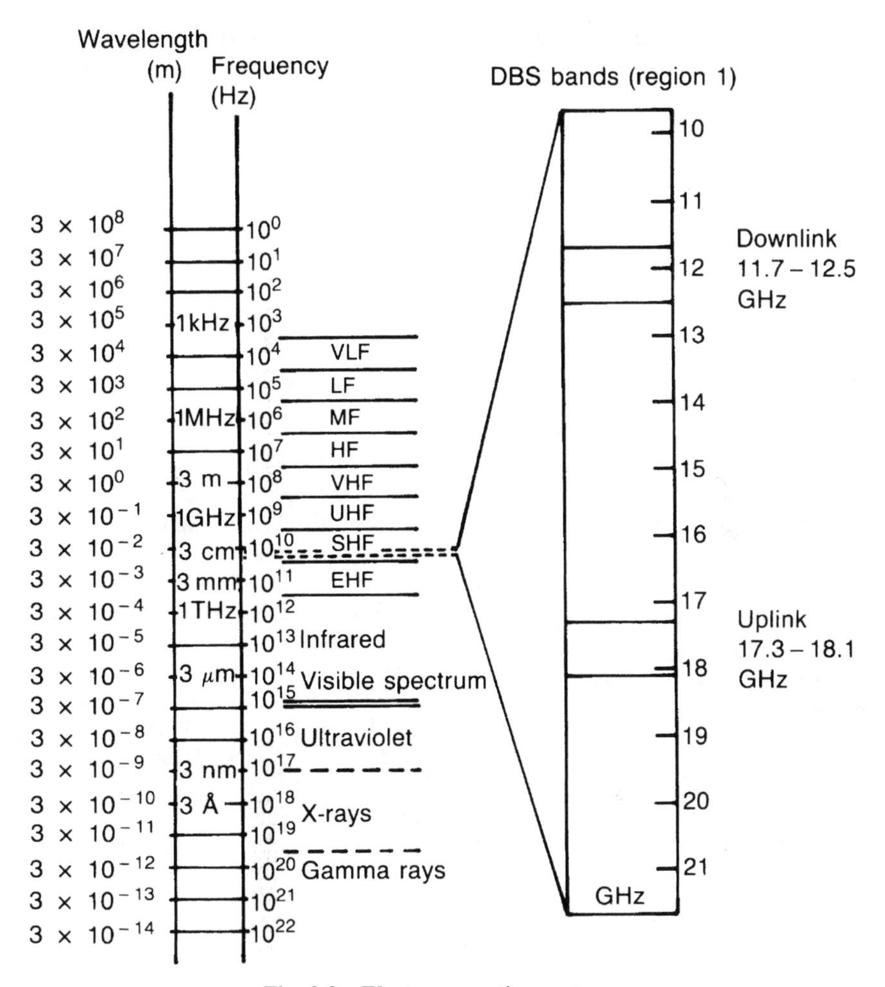

Fig. 2.9 Electromagnetic spectrum.

frequency bands to appear to have been assigned in a form of technical "one-upmanship," where each term uses an adjective that tries to outdo that of the previous name. The officially recognized terms, and the frequency bands to which they refer, are given in Table 2.3.

During World War II, for security reasons, the British and American military began to use an alternative set of terminology to refer to the specific bands that were in use for either communications or radar applications. These terms are still in common use, but not all have an officially recognized or consistent definition of exactly what frequency ranges they cover. At the lower ranges, they generally follow definitions in the IEEE Standard 521 for radar frequency bands; however, in the higher-frequency bands, this may not be the case. The approximate ranges, however, are given in Table 2.4.

Table 2.3 Official ITU frequency band designations

Name	Abbreviation	Frequency range
Very low frequency	VLF	3–30 KHz
Low frequency	LF	30–300 KHz
Medium frequency	MF	300–3000 KHz
High frequency	HF	3–30 MHz
Very high frequency	VHF	30–300 MHZ
Ultra high frequency	UHF	300–3000 MHz
Super high frequency	SHF	3–30 GHz
Extremely high frequency	EHF	30–300 GHz

The RF spectrum is a limited resource. There are far more transmitters wishing to make use of the spectrum than there are discrete frequencies available. Fortunately, however, as will be discussed later in this chapter, the antennas used on both satellites and Earth terminals have directivity. That is, they are able to transmit an RF signal preferentially in one direction, and, within limits, not transmit the same signal in other directions. A satellite can therefore transmit a number of narrow spot beams to the ground and, provided that they are sufficiently widely separated, the same carrier frequencies can be used in each beam, each carrying different content. Likewise, Earth terminal antennas can direct their transmissions preferentially toward one satellite and, provided that a second satellite is sufficiently far away from the first, the same frequency can also be used by other terminals to communicate with it. Both of these cases are examples of frequency reuse, whereby the same frequency band can be reused a number of times. In this case, the frequency reuse is provided through the means

Table 2.4 Approximate frequency ranges for unofficial band terms

Name	Approximate frequency	Typical usage
L-Band	1–2 GHz	Mobile communications, sound broadcast
S-Band	2–4 GHz	Mobile communications, sound broadcast
C-Band	4–8 GHZ	Fixed services
X-Band	8–12 GHz	Military communications
Ku-Band	12–18 GHz	Fixed services, data relay, and TV broadcast
Ka-Band	18–40 GHz	Fixed services and data relay
Q-Band	40–50 GHz	Data relay
V-Band	50–70 GHz	Data relay
W-Band	70–120 GHz	Military data relay

of geographical separation of beams, referred to as geographic discrimination, but as we will see, this is not the only way of reusing frequencies.

An electromagnetic wave consists of oscillating electric and magnetic field vectors, perpendicular to each other and each perpendicular to the direction of propagation of the wave. If an antenna produces waves with a constant orientation of these two fields in space, the wave is said to be linearly polarized (refer to Fig. 2.10). An antenna on an Earth terminal may for example, produce a modulated carrier wave whose electric field is oriented parallel to the horizontal plane. This is referred to as a horizontally polarized transmission. Another, colocated antenna can also produce a second carrier wave, traveling in the same direction but with an electric field that remains oriented in a direction perpendicular to the horizontal plane. This is referred to as a vertically polarized transmission. Receiving antennas can be made to have polarization discrimination. That is the capability to receive signals of only one polarization, rejecting those on the perpendicular polarization. Thus, we have another example where the same frequency can be used to transmit two different information signals without interference between the two signals. This is another means of frequency reuse.

In addition to linearly polarized signals, antennas can be made to produce waves whose electric fields rotate with time as they pass though space. This rotation, as seen from the antenna producing the waves, can be in either a clockwise or counterclockwise sense. These are referred to as circularly polarized transmissions, illustrated in Fig. 2.11.[9] Receive antennas can be made to accept circularly polarized signals of only one sense, and to reject that rotating in the opposite sense. Here again, we can use the same frequency to transmit two different signals in opposite senses of circular polarization without them interfering with each other, which is another way to achieve frequency reuse.

One might construe from the previous explanation that the use of polarization discrimination would allow the same frequency to be used four times (that is, two linear and two circular) with different signals, without interference. Unfortunately, this is not the case. The rotating electric field vector in a circularly polarized transmission will, at a certain time, be oriented vertically, and a quarter-period later will be oriented horizontally. It will therefore interfere with signals being received by a linearly polarized antenna. Likewise, linearly polarized signals will interfere with circularly polarized transmission. Thus, only two-fold frequency reuse can be achieved solely on the basis of polarization discrimination.

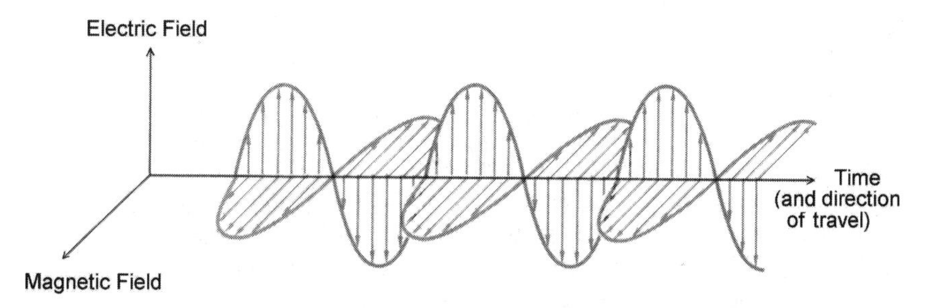

Fig. 2.10 Linearly polarized wave.

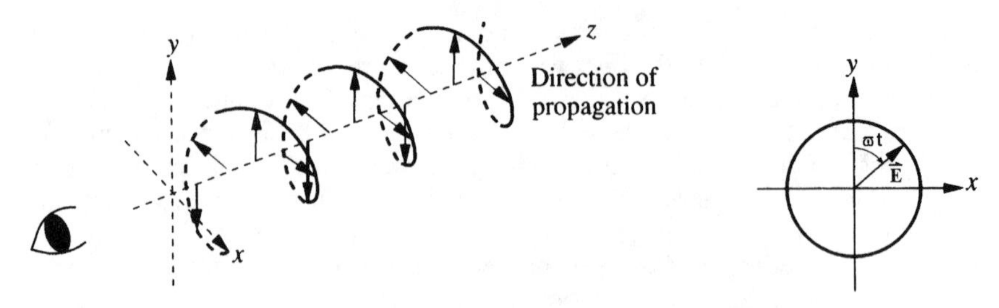

Fig. 2.11 Propagation of a circularly polarized wave.

D. Multiple Access

If two different modulated RF carriers are received, each having different information content but carriers of the same frequency, the received signals can interfere with one another. To avoid this, when a number of different terminals are using the same satellite, they can be assigned either different carrier frequencies, or, if using the same frequencies, assigned different times at which they are allowed to transmit.

If terminals are assigned different frequencies, many different transmitters can use the same satellite simultaneously. This technique is known as frequency division multiple access (FDMA), and is applicable for both analog and digital transmissions. Alternatively, a number of transmitters on the ground may be assigned the same frequency, but be allocated specific increments of time when they are allowed to transmit on that frequency. In this case, a network control center (NCC) usually sends out a common timing or clock signal to all users, so they can synchronize their transmission in time. This technique is known as time division multiple access (TDMA). The NCC divides time into periods known as frames; each frame is further subdivided into slots. Users are assigned one or more slots at fixed temporal locations within each frame when they are allowed to transmit. After transmitting a burst of (generally digital) information in their slot, they must wait until the corresponding slot(s) in the next frame interval before they can transmit again. In this way, although many users may be using the same transmission frequency, the satellite receives only one transmission at a time, and can pass it along without experiencing interference. Figure 2.12[10] illustrates this type of frequency reuse. A composite technique exists, whereby multiple transmitters are assigned both different frequencies at which to operate and discrete times at which to transmit. This is referred to as multiple frequency time division multiple access (MF-TDMA).

A third method exists for digital transmissions, in which a relatively narrow bandwidth signal is modulated with a wide band digital code, producing a wide bandwidth transmission. A second narrow band signal can be modulated with a different wide-band coded signal, and transmitted in the same wide bandwidth frequency band as the first. While these signals may and often do partially interfere with each other, they can be separated out successfully at the receiving end, provided that the codes have been suitably chosen and that the receiver knows the wide-band code that was used at the transmitter for its intended signal. This technique is known as code division multiple access (CDMA), and the transmitted

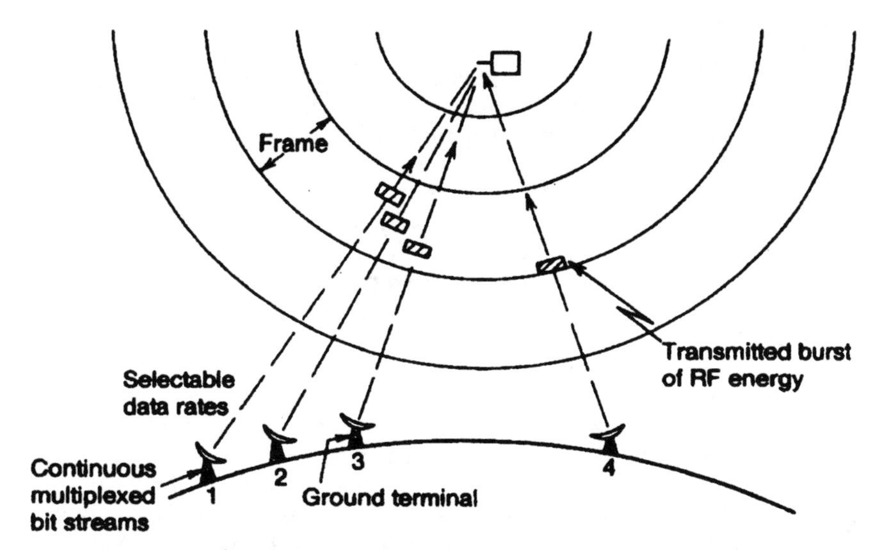

Fig. 2.12 The principle of time division multiple access (TDMA).

signals are known as spread spectrum signals. Because it allows different signals to share the same frequency band, CDMA could also be thought of as an additional means of achieving frequency reuse.

E. Compression

One of the main reasons for the evolution from analog to digital satellite transmissions is the ease with which digital transmissions can be compressed. Let us consider the case of television transmissions, for example. For terrestrial broadcasts, the standard in the United States, the National Television System Committee (NTSC) color standard calls for a bandwidth of 6 MHz for each channel. In Europe, and elsewhere that the higher-quality phase alternating by line (PAL) or sequential coleur a memoire (SECAM) standards are in use, 8 MHz are allocated to each channel.

Television images can be thought of as being composed of picture elements called pixels. Each frame of an NTSC channel consists of 488 lines that are successively "drawn" on the screen and displayed (actually, 525 lines are transmitted, but part of these are blanked out while the television is performing other functions). Thus, 488 pixels can be displayed in the vertical dimension on an NTSC television. With PAL and SECAM broadcasts, 576 active lines are displayed each frame (out of 625 transmitted); hence, the vertical resolution is 576 pixels.

Each pixel can have different levels of intensity (levels of gray in a monochrome transmission) and any one of many thousands of colors. For good picture reproduction, 8 bits per pixel are usually assigned to gray scales (yielding $2^8 = 256$ levels), and at least 16 bits/pixel to colors (providing for $2^{16} = 65{,}536$ possible different colors). NTSC frames are refreshed at the rate of 30/s, vs 25/s for PAL or SECAM. Also, the aspect ratio (ratio of horizontal to vertical dimensions) of a standard television is 4:3. Thus, if pixels were square (in fact, they need not be, so what follows is only a close approximation), to have equal horizontal and vertical resolution, the minimum number of megabits that would have to be broadcast each second would be

a) For NTSC: 488 (vertical pixels) $\times$ 488 $\times$ 4/3 (horizontal pixels) $\times$ 30 (frames/s) $\times$ 24 (bits/pixel) = 228.618 (Mbps) for NTSC and,
b) For PAL or SECAM: 576 (vertical pixels) $\times$ 576 $\times$ 4/3 (horizontal pixels) $\times$ 25 (frames/s) $\times$ 24 (bits/pixel) = 265.421 (Mbps) for PAL/SECAM.

As previously discussed, the term spectrum efficiency for digital transmissions refers to how many bits per second can be transmitted per Hertz of carrier bandwidth. This typically varies, depending on the type of modulation used, between values of from 0.83 up to a high of approximately 2.5 bps/Hz. Thus, while with analog transmissions, these amounts of pixels can be accommodated in the bandwidths of between 6 and 8 MHz, at first sight it would appear that comparable digital transmissions, at best, would require bandwidths of at least 100 MHz or more.

Why then are digital transmissions supplanting analog? There are essentially two reasons. The first is, as noted elsewhere in this chapter, that digital signals are inherently more resistant to the adverse effects of noise than are analog signals. For equal levels of radiated signal power, digital television signals can tolerate approximately an order of magnitude more noise than an analog transmission can without showing discernible degradation in the received images. The second answer is digital compression. After an analog signal has been converted to digital form, there are ways to use digital computer techniques that would not be practical for analog signals to compress the required bandwidth needed to transmit the signals by one or two orders of magnitude, to the point where the digital signals can actually be sent using far less bandwidth than that required for analog transmissions.

The details of how this level of compression can be achieved are beyond the scope of this chapter, but the principles underlying the details are rather easy to understand. Let us consider, as an example, an NTSC television transmission of a newscaster presenting the news. Each frame is an individual snapshot in time, containing 488 (vertical pixels) $\times$ 488 $\times$ 4/3 (horizontal pixels) $\times$ 24 (bits/pixel) $\times$ 1/8 (bytes/bit) = 952,576 (bytes). The next frame in the series, a snapshot taken 1/30 of a second later, will also contain the same number of bytes of data. Let us ask ourselves, however, "How many bytes in the second frame will be different from those transmitted in the first frame?" The answer is, probably not many. The background scene surrounding the head of the newscaster on the screen, and that of the desk in front of him, will probably not have changed at all. Likewise, except for the movements of the newscaster's mouth, and perhaps his eyes if he blinks, the rest of his features may not show any significant change either. Thus, the pixels in the second frame could be compared with those in the first frame, on a pixel-by-pixel basis, and only those pixels found to have changed [the delta (Δ) values], would have to be transmitted, with the transmission indicating what the changes in the different pixels were and also where in the picture the changes had occurred. This presupposes, of course, that the receiver has the buffer storage to retain the data from the first frame and use it, together with the Δ values from the second frame, to reconstruct a full image of the second frame. In this way, the amount of data that might have to be transmitted to display the second frame might be only a small fraction of that in the transmission of the first frame. Of course, if we are watching a football game where there is rapid motion of many players, or if the camera is panning across a scene, a much larger percentage of the pixels may change between successive frames, so the degree with which the transmitted data can be reduced will be less than in the case of a newscast. Nevertheless, significant reduction is possible even then.

A second method to reduce the amount of data to be transmitted makes use of the fact that many scenes have a lot of repetition of identical pixels. When you look at a photograph of, say scattered clouds in the sky on an otherwise bright and clear day, you can appreciate that the image has a large amount of redundancy in it. A large part of the scene is taken up by expanses of sky, all perhaps having the same shade of blue coloration. The scene can be thought of as being formed from a large number of individual pixels laid out in a stacked series of rows. Each pixel can be described by a sequence of digital symbols that give its location and color. There might, however, be a row of pixels in the scene that all have the same color. Then, rather than transmitting the data from each pixel, one need only transmit the data from one such pixel, together with a number indicating how many times in a row that pixel is repeated until a different color pixel is encountered.

Motion prediction is a third method to reduce the amount of data. Let us consider again the case of an airplane flying in the sky. From one frame to another, the size and shape of the airplane may remain essentially the same; only its position on the screen changes. It is possible, in such a case, to compare the two frames and detect both the amount and direction of this position change. These data can then be transmitted instead of the data of each pixel making up the picture of the airplane. At the receiver, a stored set of data for the airplane could be used, together with the direction and magnitude information, to properly position an image of the airplane on the screen. This could be continued in successive frames until there is a change detected, at the transmitter, in the size or aspect angle of the airplane. When this occurs, new data describing each pixel making up the picture of the airplane will again have to be transmitted, but this may not have to occur until many successive frames have gone by with only a minimal amount of data having to be sent in the meantime.

Finally, the human eye is not equally sensitive to all colors and to all changes in color or intensity between adjacent colors. It is possible, therefore, to eliminate certain data that are below the threshold of detection of the eye, or that are only slightly above this threshold, without significantly reducing the perceived quality of the picture. Small differences in colors and intensities that are least apparent to the eye can also be neglected.

After a scene has been converted from its native analog form to digital, any or all of the techniques previously described can be used to reduce the amount of data that needs to be transmitted to receive and recreate the picture. A number of internationally accepted standards have been developed to specify how this process should be done. The most commonly used standard today is called the

MPEG-2 (Motion Picture Experts Group standard version 2). Using the MPEG-2 techniques, the amount of data that needs to be transmitted to obtain an acceptable quality television picture can be reduced from the 100 Mbps or higher values previously discussed to only a small percentage of this value. The level of such compression that can be reached depends on both the level of quality demanded in the received signal and the type of program being transmitted. For example, a typical VHS-quality video can be compressed and broadcast at 1.5–2.0 Mbps, a standard quality television at approximately 4.0 Mbps, live-action sports at approximately 6.0 Mbps, and studio-quality broadcasts at 8.0 Mbps (Ref. 7).

MPEG-2 compression is the standard in use at the beginning of the 21st century for digital television broadcasting via satellite. A combination of high-speed computers plus special-to-type hardware circuitry allows real-time compression and broadcasting of television programming. The compression achieved allows as many as eight channels of television to be transmitted in the bandwidth previously required for transmission of a single analog channel. The MPEG standards, however, are continuing to evolve. MPEG-4 now exists, which can offer even higher levels of compression, as well as improvements in quality over MPEG-2. The encoding process with MPEG-4 is, however, considerably more complex than that of MPEG-2. As of this writing, the computations required for MPEG-4 compression and encoding are too complicated and lengthy to be done in real time. They must be done off-line, with the process taking considerably longer than that for MPEG-2. Decoding, however, can be done relatively quickly; therefore, real-time viewing of streaming MPEG-4 data is no particular problem.

F. Frequency Conversion

As mentioned previously, the satellite payload cannot retransmit signals down to the ground at the same frequency as the signals it receives from the Earth station. The uplink frequency therefore has to be shifted to another frequency band before it is transmitted down to the ground. Higher frequencies are attenuated more than lower ones by weather conditions in the atmosphere, and this attenuation requires a transmitter to be able to radiate extra power to overcome it. Because extra power is far easier to come by on the ground than in space, normally the signals sent to the satellite from the Earth station are chosen to be at a higher frequency than that of the return signals from the satellite back down to the ground. The signals received at the satellite, therefore, generally must be converted to a lower frequency band through the use of a downconverter before being retransmitted.

A downconverter makes use of a characteristic of nonlinear electronic devices. A linear amplifier, for example, takes an input signal, boosts its amplitude by a fixed gain, K_A, and outputs the boosted signal. In a nonlinear amplifier, the value of K_A is not a constant, but instead is a variable whose value depends on the magnitude of the input signal itself.

If two separate frequencies, f_1 and f_2 are sent simultaneously through a linear amplifier, the output will consist of a composite signal containing just these two frequencies. If the same frequencies are fed to a nonlinear amplifier (or indeed, to any nonlinear device) the output will consist of these two frequencies plus a signal whose frequency is equal to the sum of the two frequencies, plus a signal whose frequency is equal to the difference of the two frequencies, (plus, in general, an infinite number of higher-order terms whose frequencies are sums and differences of integral multiples of each of the input frequencies, but with decreasing amplitudes). It is then possible, using a band-pass filter, to separate out a specific one of the infinite number of frequencies present at the output of the nonlinear device. In a downconverter, the filter is chosen to single out only that term whose frequency is equal to the difference of the two input frequencies. One of the inputs, for example, could be a received composite television signal with a frequency in the band 17.3–18.1 GHz band [allocated for uplinks for direct broadcast satellites (DBS) in Europe], and the other could be a fixed frequency of 5.6 GHz, generated on board the satellite by a local oscillator (LO). The output signal will then contain elements having a difference frequency in the band 11.7–12.5 GHz (allocated to downlinks in Europe for DBS), which can be separated out with a suitable band-pass filter.

An upconverter may also be needed for some types of payloads (described later in this chapter). This works just like a downconverter, but uses a value of LO frequency and a filter (usually selected to obtain the term whose frequency is the sum of the IF and LO frequencies).

G. Antennas

Antennas are perhaps the most important element in the entire transmission chain because they provide gain and discrimination that determines whether a satellite system will work at all in the presence of noise and interference from other sources, particularly when there is more than a single satellite in orbit. Antennas, therefore, deserve a more in-depth explanation than the other elements described in this chapter. Also, despite promising that the use of equations would be minimal in this book, it will be necessary to include a number in this section because these are essential for an understanding of how antennas operate and why they are so important.

As mentioned previously, antennas exhibit directivity, allowing them, within limits, to radiate preferentially in one direction rather than in another. The two types of antenna most often used in satellite communications to provide this directivity are horn antennas and parabolic reflector antennas. To understand how either of these works to radiate a signal away from the antenna, we must first understand how signals are carried to an antenna in the first place.

Electrical wires are suitable for conducting alternating current from place to place—in telephone or power networks on the ground. The frequency of the alternating current in power systems is typically 50 or 60 Hz, and rarely more than 400 Hz. Telephone voice lines in the home carry signals in the low kHz range. At these frequencies, wires can conduct electrical currents relatively efficiently. Losses are, of course, present because of the finite electrical resistance of the wiring that converts some of the current into heat, but these losses are relatively small and can be minimized by increasing the size of the wiring used.

As the frequency of the conducted signal increases to higher than these levels, however, conventional wiring becomes increasingly less and less efficient. Two phenomena begin to play a major role in this. First, at higher frequencies, electrical impulses do not penetrate well into metallic conductors, and even with thick cables the currents begin to stay concentrated near the surface of the wires. As less of the wire's cross-sectional area becomes used to carry the current, the resistance increases, creating greater losses in the form of heat. Second, as the frequency increases, the wires begin to radiate energy, rather than conducting it, and in this case the radiated energy is lost.

Up to certain limits, coaxial cables can be used to conduct signals at high kilohertz frequencies, and even well into the megahertz range. Coaxial cables consist of a center conductor, surrounded by a layer of dielectric material, which is then encapsulated in a grounded layer of conducting material that serves to reduce the losses because of radiation. Eventually, however, frequency levels are reached where even coaxial wiring becomes too lossy, and waveguides must be used.

A waveguide is a hollow tube, typically of circular or rectangular cross section, within which very high frequency waves can be carried by reflection rather than by conduction. Electromagnetic waves propagate down a waveguide by a series of zigzag reflections from its interior walls. The dimensions of the waveguide must be chosen to be compatible with the range of frequencies that they are to carry, but provided that this is done properly, waveguides can transport electromagnetic waves with high efficiency.

Because the waves are bouncing back and forth between the walls as they pass through the waveguide, the apparent speed at which they are propagated down it is less than the speed of light at which electromagnetic radiation typically travels. When such a wave reaches the end of a waveguide, however, there is a sudden discontinuity in the speed at which the wave can propagate. This discontinuity causes part of the energy of the wave to be reflected backward and the remainder of the energy to be radiated forward into space. The radiated energy may be the desired effect, if this is to serve as an antenna itself, but the energy radiated backward is effectively lost and reduces the efficiency of a waveguide as a radiating element.

If, instead of suddenly terminating, the waveguide dimensions gradually increase starting at some point, forming a horn shape, the degree of discontinuity at the exit point is minimized, and the amount of energy reflected backward by it can be reduced greatly (see Fig. 2.13 and Ref. 9). Furthermore, with such a horn-shaped antenna, the radiated energy at the exit point can be concentrated in a preferred direction rather than radiating in many directions. The degree of this concentration, relative to one radiating equally in all directions, is referred to as the directivity of the antenna. Directive antennas are thus to be distinguished from omnidirectional antennas, which transmit radiation isotropically, that is, equally in all directions.

Consider for a moment, an isotropic antenna at the center of an imaginary sphere of radius R meters. If the antenna is radiating a power of P_t W, and if the space inside the sphere is a perfect vacuum so that it does not absorb electromagnetic energy, then the radiated power will all pass through the surface of the sphere and will be spread out evenly over the surface. Because the surface area of a sphere is $4\pi R^2$, the flux density (that is, power per unit area) at the surface of the sphere will be given by

$$F = \frac{P_t}{4\pi R^2} \text{ in W/m}^2 \tag{2.4}$$

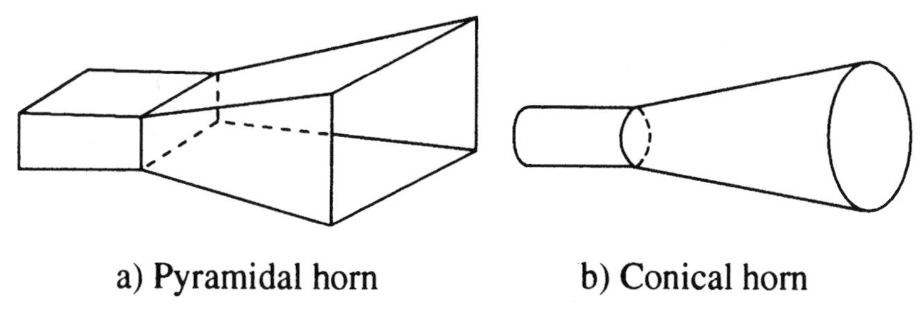

Fig. 2.13 Waveguides feeding horn antennas.

In other words, isotropic radiation follows what is called the inverse square law with distance, just as do gravitational, electrostatic, and magnetic fields. The strength of each such field varies inversely as the square of the distance from their source.

Directive antennas can be compared in performance to omnidirectional antennas in terms of their gain. This can be defined as the ratio of the power flux density from the directive antenna, measured at a certain location in the direction of maximum transmission, to the flux density that would exist at that same point if it were being transmitted from a nondirective (isotropic) antenna. If we define the term G_t to be the gain of such a directive antenna, then the flux density at any point a distance R from the antenna, measured in its direction of maximum transmission, which we will call F_d, can be written as the product of the flux density that would exist there from an isotropic antenna and the gain of the directive antenna.

$$F_d = \frac{P_t G_t}{4\pi R^2} \text{ in W/m}^2 \tag{2.5}$$

The term $P_t G_t$ in this equation is defined as the equivalent isotropic radiated power (EIRP), and is one of the most common (and often, least understood) terms used to compare different satellites in the jargon of satellite communications. In simplest terms, however, the higher a satellite's EIRP, the easier it will be for receivers on the ground to detect its signals without interference from other signals or noise sources. From its definition, one can see that a satellite designer has two ways to increase a satellite's EIRP—either increase the level of the output power radiated or increase the gain of the radiating antenna. Because satellite power has to come from either solar arrays or batteries, both of which are very expensive to build and put into space, great emphasis is put on making a satellite's antennas as high performing as possible.

The directivity that can be obtained from a practical horn antenna is limited. To go above its limits, the most common way to improve the directivity of a satellite antenna is to use a parabolic reflector to concentrate radiated energy in the preferred direction. A parabolic reflector (or more precisely, a paraboloid of revolution) has the characteristic that rays originating at a point called its focus will all be reflected from its surface in the same direction, forming a parallel beam. This is depicted in Fig. 2.14.

If a horn as shown in Fig. 2.13 is placed at the focal point, F, of an ideal parabolic reflector, and radiates into the reflector, the energy in the beam radiated

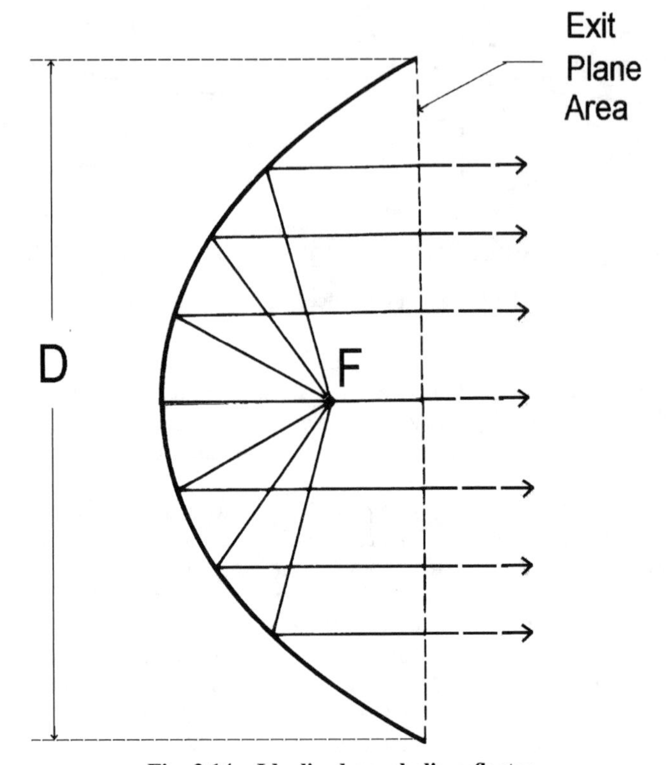

Fig. 2.14 Idealized parabolic reflector.

from the reflector might then be thought to all be concentrated into a single beam pointing in a single direction. Unfortunately, with electromagnetic radiation as with many other things, the world doesn't work in an ideal manner, so the strict unidirectional character of the output beam shown in the figure is not obtainable in real life. In the first place, for a satellite communications antenna, the rays do not originate from an ideal point *F* as in Fig. 2.14. The feed (typically some form of antenna horn as was previously described) is not all at the single point *F*. Because the exit area of the horn has finite dimensions, most rays originate from only near the focus, not from an ideal single point, and hence are not all reflected in the same direction. In the second place, electromagnetic radiation is not only subject to reflection, as illustrated in the Fig. 2.14, but also to diffraction.

If a parallel beam of monochromatic (that is, all of a single color) light passes through a large circular hole and then impinges on a screen some distance behind the hole, the image on the screen will be that of a circle, of the same size as the hole. If the size of the hole is decreased, the size of the circular image will also decrease. If this is continued, however, there comes a point when the size of the hole begins to become comparable in order of magnitude to the wavelength of the light in the beam, where one begins to get, not a circular image, but rather a series of concentric rings. This phenomenon, because of the constructive and destructive interference of diffracted light, is depicted in Fig. 2.15.

A similar thing happens with antennas and RF radiation. An antenna produces a large primary beam in one preferential direction (called the boresight direction) and a series of decreasingly weaker beams (called sidelobes) at increasing angles away from the boresight direction. A two-dimensional cross section of such a multiple beam pattern is shown in Fig. 2.16.

What Fig. 2.16 shows is that the majority of the energy radiated from the antenna is spread out over a certain angle into a beam known as the main beam or primary lobe, and that some small portions of it are also radiated at relatively large angles away from the boresight into the sidelobes. This means that an antenna on the ground that is radiating a signal to a satellite in orbit may also be sending part of its signal to nearby satellites and causing interference to them. This, however, is not the end of the story. Antennas are, in general, symmetrical in their performance. That is, they can act not only as transmitters but also as receivers, and their behavior and performance in either direction will be the same as long as the frequency of operation is the same. Thus, receive antennas also have patterns exhibiting side lobes, and the antennas on a satellite that are designed to receive signals from a single location on Earth can also receive signals from other terrestrial transmitters located nearby, as well as from more distant transmitters if their locations correspond to one of the receive sidelobes.

Fortunately, there are ways to reduce the possibilities of interference caused by (or caused to) other systems of transmitters or receivers. Perhaps the most effective way is to make the gain of the antenna as large as possible in the boresight direction, make the beamwidth of the primary lobe as narrow as possible, and make the magnitude of the sidelobes (that is, the gain in the direction of each sidelobe) as small as possible.

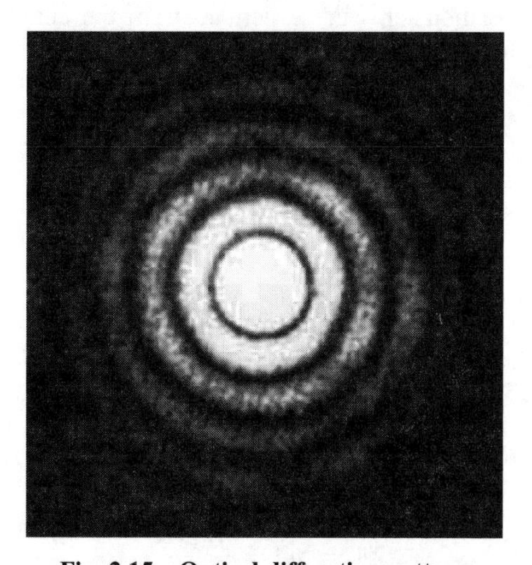

Fig. 2.15 Optical diffraction pattern.

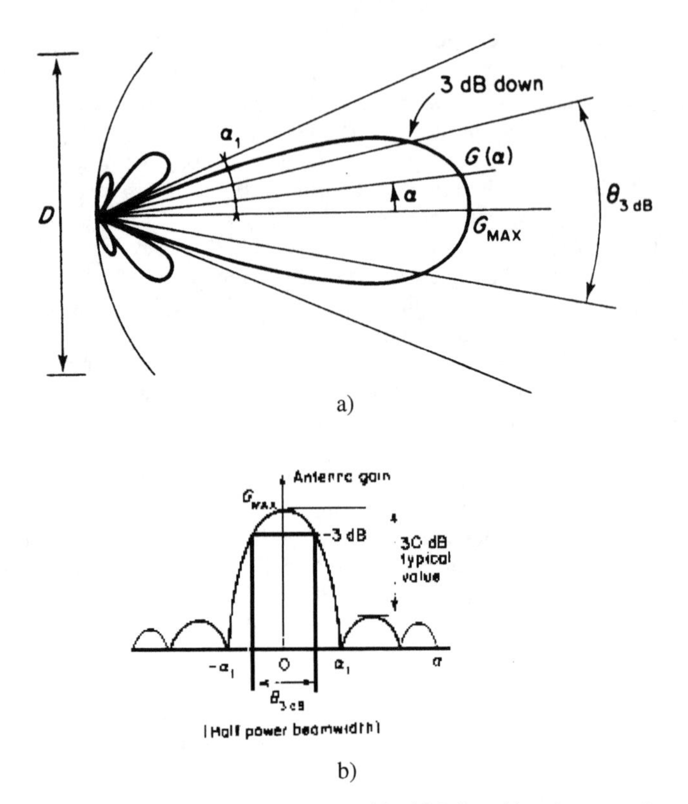

Fig. 2.16 a) Antenna patterns with sidelobes, b) gain vs angle.

A fundamental relationship can be derived from the antenna theory, which shows that the maximum gain, $G_{\max}$, of an antenna (be it a horn or a reflector type) is related to the wavelength of the signal being transmitted and to the effective cross-sectional area of the antenna's exit plane

$$G_{\max} = \frac{4\pi A_e}{\lambda^2} \tag{2.6}$$

In this equation, the effective area of an antenna is somewhat less than its actual exit area shown in Fig. 2.14 because of various inefficiencies inherent in any real antenna (for example, because of manufacturing tolerances or blockages due to the feed horn and its supports). If we use the symbol "η" to indicate the overall efficiency of the antenna, (η typically lies in the range of between 0.55 and 0.70 according to Ref. 11), then the effective area can be written in terms of the actual exit area as

$$A_e = \eta A = \frac{\eta \pi D^2}{4} \tag{2.7}$$

Substituting this equation into Eq. (2.5) yields the relationship

$$G_{\max} = \eta\left(\frac{\pi D}{\lambda}\right)^2 \tag{2.8}$$

From this, one can see that for a fixed wavelength the best way to increase the gain of an antenna is to increase its diameter. Improvements in efficiency will also help, but because the efficiency cannot have a value greater than 1.0, there is a limit to how great an improvement can be made by that means only.

The width of the main beam is normally defined in terms of its half-power beamwidth. That is the angle, measured from one side of the boresight direction to the other, inside of which half of the power radiated from the antenna into the main beam lies, with the other half lying outside that angle. This is labeled as Θ_{3dB} in Fig. 2.16, because power in a communications system is normally measured in decibel watts rather than absolute watts. A decibel watt, dBW, is defined as ten times the logarithm of the power in watts. This angle depends on several of the same factors that determine an antenna's efficiency, and in particular on just how the energy, radiated into the reflector from the horn feed, is spread out over the reflector's surface (referred to as the aperture illumination). In general, however, the half-power beamwidth can be written as:

$$\Theta_{3dB} = \frac{\alpha\lambda}{D} \text{ (in degrees)} \tag{2.9}$$

where α may range between approximately 58 and 70, depending on the type of aperture illumination.[11] For typical satellite antennas, the higher value is usually used because it relates to the type of aperture illumination that also minimizes the magnitude of sidelobes.

We thus see that increasing the diameter of the antenna reflector both increases the gain in the boresight direction and decreases the width of the main beam. There are, however, physical and practical limits to how much a satellite antenna reflector's diameter can be increased. First, satellites, including their antennas, have to be launched on rockets whose shrouds (the protective enclosure on top of a launcher inside which the satellite to be launched is placed) have fixed diameters. Reflectors are typically folded against the sides or top of a satellite during launch, but their maximum dimensions must still fit inside. While it is possible to build reflectors that can themselves be folded up in some way to fit inside a shroud, and then to deploy these in orbit after the satellite has been released from the launcher, it becomes increasingly difficult to do this with sufficient accuracy at the higher frequency levels used by many satellites. Thus, designers try to keep the antenna sizes small enough to fit inside the shroud of the launcher to be used.

The second limitation comes about from the narrowing of the beamwidth as the diameter of the reflector is increased. There are limits as to how accurately the attitude control system on a communications satellite can aim and maintain the pointing of the boresight direction of an antenna. This typically cannot be done much better than to within 0.05 to 0.10 deg. If the beamwidth of the antenna gets much narrower than approximately five times the accuracy of pointing, too much of the radiated power may be directed away from where it is needed to serve the users of the satellite. Thus, satellite designers usually avoid beamwidths of less than approximately 0.5 deg.

While a parabolic reflector has only one geometric focal point, it is possible to place a number of feeds close to the focal point, and each will form a discrete beam, pointing in slightly different directions. The individual beams generated from feeds that are not exactly at the focus will be a bit distorted compared to a focal-point-fed beam, but these will still be useful for connecting with or broadcasting to different regions on the ground. If a group of feeds are driven by the same source signal, a shaped coverage can be achieved. If the feeds are independently driven, this is an example of a multibeam antenna, where a number of spot beams are produced from a single reflector.

There is a third type of antenna that is in use on some satellites, called a phased-array antenna. This consists of a number of very small antennas, called elements, packed closely together, with each antenna being fed the same signal but with very small and precisely calculated differences in the phase angles at which the signals are fed to each antenna. The result is a beam that can point in any of many different directions, depending on the phase differences used.

Explaining in detail the way this works is beyond the scope of this book. In principle, however, it works using the principle of diffraction previously discussed, where signals from different sources can interfere with or reinforce the other's signals. Suppose we have two small antennas, A and B, located a distance D apart, radiating omnidirectionally. If the same signal is sent to both antennas, but the signal to B is delayed with respect to that going to A, then the signals from the two antennas can reinforce each other in a direction that is deflected from the direction of their respective boresight. If many antenna elements are used, and they are in a two-dimensional array rather than in the linear array, then the beam can be oriented in two dimensions. Furthermore, if several signals of different frequencies and phases are fed to the multiple elements, it is possible to form several beams, each pointing in a different direction if desired, from the same array of elements.

As was the case for the horn or reflector, array antennas follow the same general rule. The total area encompassing the many elements in a phased array determines both the array gain and the width of its beam(s). The total number of elements packed into this area will influence how homogeneous the beam is and the efficiency of the antenna.

Finally, it should be noted that combinations of the various types of antennas previously described may be used in particular instances. For example, one could use a phased array to feed a large reflector.

V. Satellite Link Elements

A. Earth Stations

Previously discussed, a satellite link consists of a transmitter, a transmission medium, and a receiver. Let us examine each of these in some detail.

An Earth station transmitter is designed to accept information signals of whatever kind (including voice signals, radio or television broadcasts, and data), use them to modulate a carrier, possibly add error detection or correction coding, and amplify and radiate the modulated signal toward a satellite. Figure 2.17 shows

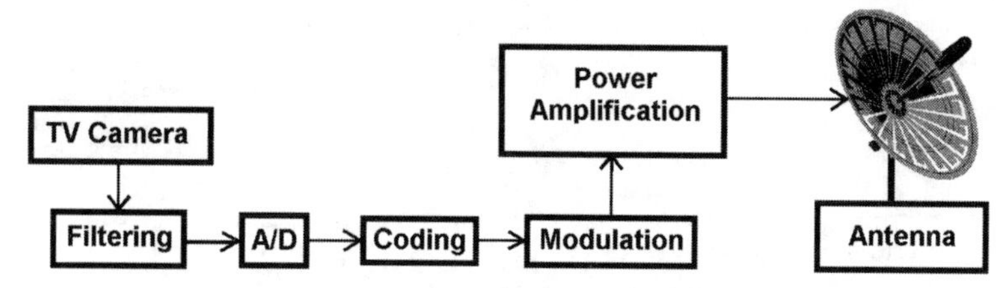

Fig. 2.17 Typical Earth station transmitter elements.

a simple block diagram of the main elements involved in such a transmitter. This shows, for example, the elements that might constitute a satellite news gathering (SNG) uplink station.

The first item in Fig. 2.17 is the TV camera, which produces an analog television signal. This is then fed to a filter, then into an analog to digital converter (A/D). The coding is then added to the digital signal coming out of the A/D, which is then fed to the modulator where it is combined with the high-frequency carrier to produce the modulated signal. That signal is amplified before being fed to the antenna, which radiates the signal toward a satellite in orbit.

The purpose of coding and modulation have already been discussed, and the TV camera is only given as an example of the many types of analog signal source that might produce the signals that are to be sent via satellite. The other elements in Fig. 2.17, however, require some explanation as to what they are and what their purpose is in the Earth station.

There is a theorem, called the Nyquist Sampling Theorem,[8] which states that if an analog waveform is sampled with a sampling frequency at least twice the frequency of the highest frequency present in the waveform, the analog waveform can be reconstructed exactly from the sampled data points. However, if there are frequencies present in the waveform that are higher than half the sampling frequency, these will cause a distortion, known as aliasing, when attempting to reconstruct the original waveform from the sampled data points.

To avoid aliasing, the TV signal to be sampled in the example in Fig. 2.17 is first passed through a filter that greatly attenuates or eliminates frequencies above a certain frequency. The waveform exiting the filter can be thought of as a truncated Fourier series that still is as close as desired to the original TV signal. It is then sampled in the A/D conversion at a frequency a little over twice as high as the highest frequency that can pass through the filter. The resulting digital data stream, when it finally reaches the receiver, can then be converted back into the waveform that exited the filter without encountering aliasing problems (see Fig. 2.18).

The modulated signal exiting the modulator is a low-power signal that must be amplified to provide a sufficient power level before being fed to the antenna. This power amplification can be provided either by high-power transmitting tubes, or by high-power, solid-state amplifiers. The current state of the art, however, predicates the use of amplifying tubes if more than a few tens of watts of output power are required. After amplification by either method, however, the resulting high-power signal is fed to the transmitting antenna, whose operation has been previously explained.

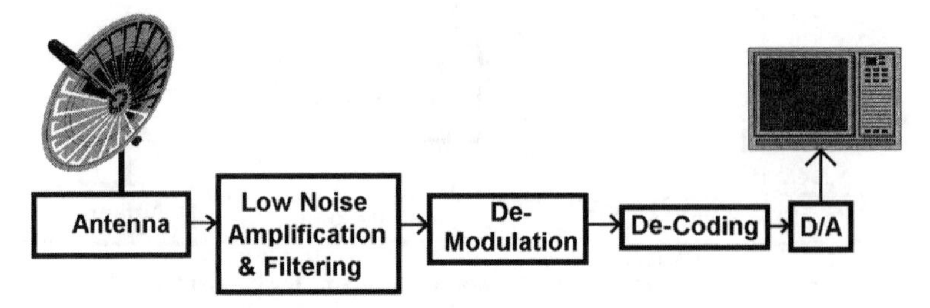

Fig. 2.18 Typical receive Earth station.

In the previous paragraphs, we have described the major elements that make up a simple satellite transmitting Earth station. In practice, Earth stations do not have only a single source of content. Often, for example, multiple television programs are fed to the Earth station, which must multiplex these into a single uplink transmission to a satellite. The multiplexing, particular of digital signals, is often done in an analogous manner to the TDMA multiple access methods previously described. In this case, however, instead of multiple Earth stations being assigned time slots during which they can broadcast information, a single Earth station assigns each television signal a time slot, combines several of these into a continuous uplink beam, and transmits this to the satellite. This type of uplink is called a time division multiplex (TDM) bit stream.

B. Satellite Repeater

The Earth station forms one end of a satellite link, while the communications payload on the satellite forms the other. Many of the elements that make up the satellite payload are similar to those that we have described for the ground station, as we shall see, when describing the elements in Fig. 2.19).[8]

A typical communications/broadcasting satellite payload consists of one or more antennas together with one or more repeaters. The antennas receive low-level signals from a particular direction and radiate high-power signals toward a desired direction. The repeater performs low-noise amplification, carries out frequency conversions, typically channelizes signals, and provides power amplification.

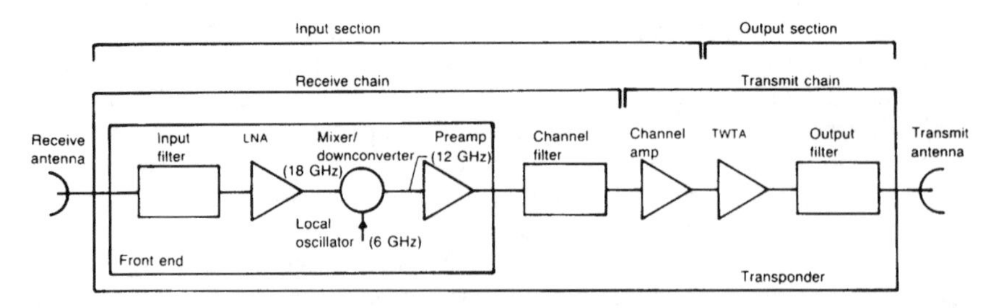

Fig. 2.19 Single conversion satellite repeater.

What are the functions that the communications payload must perform? Even with an antenna with very high gain and efficiency at the Earth station, the radiated signal spreads out and dissipates in space, and only a tiny fraction of the power transmitted actually reaches a satellite. The signal reaching the satellite antennas must then be highly amplified before it becomes useful. The signal received by a satellite antenna is therefore first fed to a low-noise amplifier, or LNA (that is, one that does not add much additional electrical noise to the signal). Because the antenna may also be receiving signals at different frequencies from Earth stations operating with other satellites, the LNA usually includes a band-pass filter that eliminates all frequencies outside those of interest at the moment.

There are two ways in common use to carry out the frequency conversion. A downconverter can either convert the received and amplified signals down directly to the frequency band to be used for transmission (called single conversion), or it can convert the received frequency down to a much lower frequency range, called an intermediate frequency (IF), further amplify and/or filter the IF signals, and then an upconverter can boost the signal up again to the desired transmit frequency band.

This dual conversion shown in Fig. 2.20[8] may seem overly complicated, but it can making certain filtering or switching operations much easier to perform, and is a means to avoid certain frequencies that might otherwise cause internal interference; therefore, the additional complexity is often justified. Furthermore, in OBP payloads signals are normally down converted to a low IF or even to baseband before being processed, making double conversion a necessity in this case.

If the signal contains signals from multiple sources (as in an FDMA system), or is very broadband in nature, then after downconversion it will typically be fed to a bank of filters that channellize the broadband composite signal into a series of narrower band frequencies. These are then individually amplified and further filtered before the individual signals are either individually transmitted back to the ground or recombined again in what is called an output multiplexer (omux) and then transmitted as a composite beam down to Earth. There are several reasons why such channelization might be done, but one of the more important is that the separate user signals do not typically each occupy the entire band. In addition, it is difficult to make solid-state amplifiers that have both a constant gain over a wide frequency range and a relatively high efficiency. Channelization slices a broad input frequency band into multiple signals, each having a relatively narrow band of frequencies, for which solid-state amplification is not such a problem. The final stage of power amplification before feeding the signal to the downlink antenna is usually provided by one or

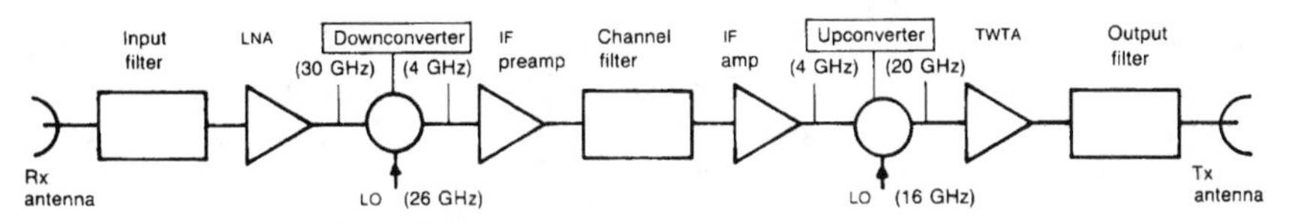

Fig. 2.20 Dual conversion repeater.

more traveling wave tube amplifiers (TWTA), which are inherently broader band amplifier types. The channelized signals may be recombined into a single broadbeam before being fed to the TWTAs.

C. Onboard Processing

The repeater technology described thus far has been of the bent-pipe variety. Signals have gone up in one beam, been received, and the frequency has been shifted, amplified, and broadcast down again in a single beam. What happens, however, if we have a multibeam system where a number of narrow spotbeams cover an area, rather than a single beam? Signals can come up in any beam from terminals located within its coverage, but might be directed toward another user that is in the coverage of another beam. How can we direct such signals toward their intended destination beam?

If we wish to remain with a bent-pipe type of payload, there are basically two ways this can be accomplished. First, we can do it by the use of frequency discrimination. Suppose we have 8 spot beams, numbered 1 to 8, each of which is assigned a band of frequencies 80-MHz wide (therefore we use a total of 640 MHz of spectrum). We could decide that, for each uplink beam, this 80 MHz is to be divided up into 8 blocks of 10 MHz each. Within each beam, we would assign filters and connections to route any signal sent on the uplink received in the lowest 10-MHz block automatically to the downlink of beam 1. If a signal is received in the next to lowest block, it will be routed to the downlink of beam 2, and so on for all 8 beams. Then, all that is necessary is for a terminal to have the capability of sending its information into any of the 12 blocks to ensure that the signal is sent to the proper downlink beam.

This is relatively easy to implement, but it would lead to highly inefficient use of the scarce spectrum resource if, as would almost certainly be the case, the density of traffic to and from each spot beam was not the same. It would be difficult to predict just what the traffic per beam would be before launch, and because the typical lifetime of a satellite in orbit might be 15 years or more, it certainly would not be possible to predict that far in advance. What if the number of users in a particular beam needs more than 10 MHz at any one time? Many might have to wait an objectionable period before being able to get their message through. In addition, one of the purposes of using spot beams in the first place is to reuse frequencies between different beams, which would complicate any system trying to interconnect any spot beam to any other in this manner.

What bent-pipe, multibeam system designers contemplate doing, therefore, for such a system, is to group the 8 beams into smaller groups, for example, 2 groups of 4 beams. Furthermore, the designer would arrange the beams so that the individual spot beams in each group of four are sufficiently far away from those in the other group that the frequencies can be reused twice over. Then, each set of four beams can be assigned 640 MHz, with 160 MHz allocated to each beam. The satellite will then consolidate all the traffic received in each set of four beams, multiplex it together and send the composite beam down to a gateway station on the ground. There, the signals would be separated out and demodulated down to baseband. The header of each digital signal would then

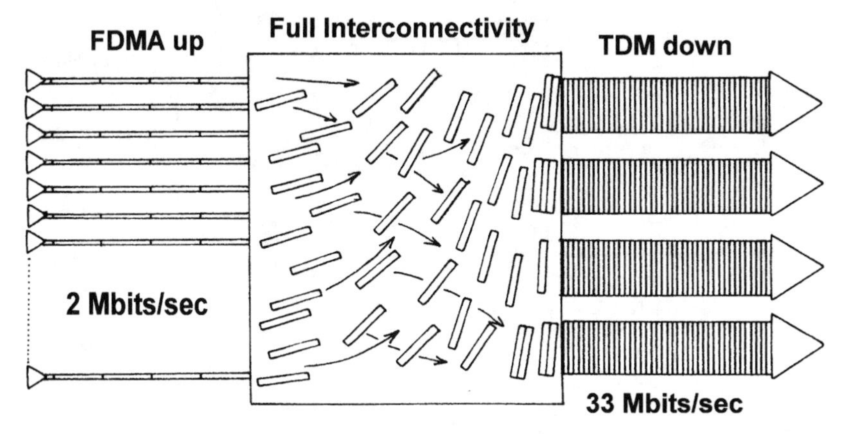

Fig. 2.21 OBP principles.

be examined by a computer, which would allow it to determine to which beam the signal was intended. The gateway computer would then frequency shift each of the signals accordingly, remodulate the signal, and either send it back up to the satellite for distribution in one of the same set of 4 beams, or send it via a terrestrial fiber link to another gateway station, servicing the other set of 4 beams.

The main problems with the gateway routing approach are that

1) It requires broad bands of frequency for both the downlink to the gateway and to the uplink back to the satellite.
2) Even at the speed of light, the "double hop" that such a process involves introduces a delay of 0.5 s or more between the transmission of a signal to the satellite and its reception at the intended terminal on the ground. This can make telephone calls or videoconferences very irritating to the participants if these are carried over such a double set of links.

To overcome these problems, an OBP payload can be used. This essentially performs the computation task on board the satellite that was previously described for a gateway station. Signals arriving at the satellite from multiple sources in a multibeam environment, using FDMA for example, can be demodulated on the satellite and then redirected to other beams, reformatted perhaps into a TDM format. Figure 2.21* shows this in pictorial form.

Onboard processing can, however, do more than Fig. 2.21 might imply. Suppose terminals on the ground might wish to access a satellite at a variety of different bit rates. An OBP processor can, in such a case, reformat these into not only different downlink beams, but also into different bit rate signals. Figure 2.22* illustrates this flexibility.

---

*Internal communications of the European Space Agency, with permission of the author, Simon Dinwiddy.

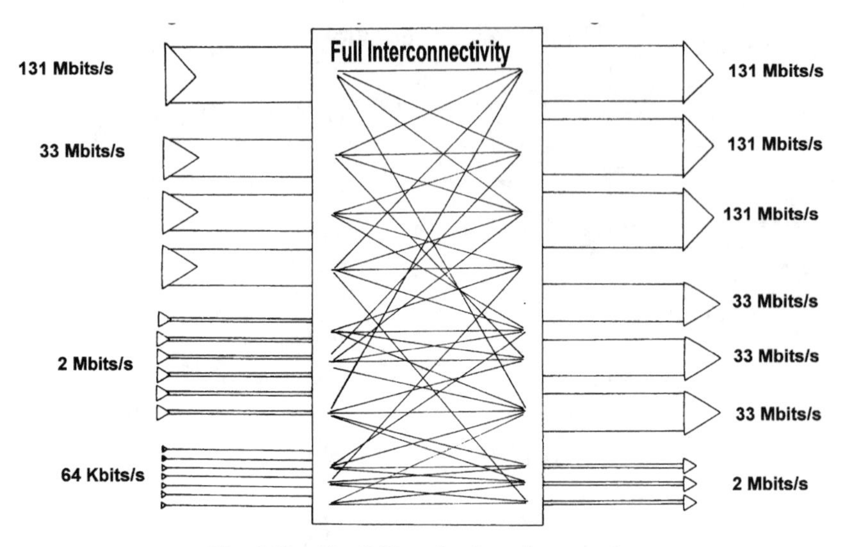

Fig. 2.22 Flexibility of onboard processing.

There is another important reason why one might wish to use an OBP rather than a bent-pipe payload. That reason is noise. A signal sent from an Earth terminal to a satellite inherently arrives at its destination "contaminated" with interference from other sources and from thermally produced noise. In a bent-pipe payload, the signal is amplified before being retransmitted back to Earth, but so is the noise. What ultimately arrives at the receiving terminal then includes noise and interference from both the uplink and the downlink.

With an OBP payload, the demodulation occurs on the satellite, and the FEC coding on the uplink signal can be used to eliminate the majority of any errors caused by interference or noise on the uplink. The original signal is therefore said to be regenerated. The receiver on the ground therefore, for the most part, only has to contend with the interference and noise of the downlink, whose correction can be aided by the addition on board the satellite of further FEC coding on the regenerated signal before it is transmitted.

D. The Link Budget

The last concept to be addressed in this chapter is that of a link budget. In the up and down links of a satellite transmission, as well as within the satellite payload itself, the power levels of signal (and noise) can vary over many orders of magnitude. A link budget is a tabulation of the signal amplifications and attenuations and losses that occur at the various points in such a dual link, which ultimately can show whether a signal transmitted from an Earth terminal can or cannot be successfully received at the other end of the dual link. Calculating a link budget is relatively straightforward, but we will have to go back to the use of equations to illustrate this. Equation (2.4) gave the flux density at a distance R from a transmitting antenna. If we place a receiving antenna at that point having an effective aperture area of A_{eff} The received power will be:

$$P_r = \left\{\frac{P_t G_t}{4\pi R^2}\right\} A_{\text{eff}} \tag{2.10}$$

Communications engineers prefer working with the gain of an antenna rather than with its effective area. From Eq. (2.6), however, we can write the effective area of the receiving antenna in terms of its gain, and substitute into Eq. (2.10) to obtain

$$P_r = \left\{\frac{P_t G_t}{4\pi R^2}\right\} \left\{\frac{G_r \lambda^2}{4\pi}\right\} \tag{2.11}$$

Equation (2.11) is usually written in the form

$$P_r = \frac{P_t G_t G_r}{\{4\pi R/\lambda\}^2} \tag{2.12}$$

The term in the denominator is referred to in the jargon of communications as the free space loss and denoted as L. The name, however, is a misnomer because empty space does not absorb energy, so nothing is actually lost. The R part of L simply reflects the fact that radiation follows an inverse square law, but the "λ" part has nothing to do with either this or any kind of real loss. It appears in the equation only because we have written the equation in terms of the receive antenna gain, rather than simply its effective area.

In the discussion concerning Eq. (2.9), the concept of the decibel watt was introduced. To understand how a link budget is normally computed, this concept may require a bit of elaboration for those readers who may not be familiar with the term.

In nonrigorous terms, the decibel value of a quantity is defined as ten times the logarithm to the base ten of that quantity. Thus W watts, expressed in decibels (written dB), is given by $10\log_{10}\{W\}$ dBW. However, mathematicians object rigorously to taking the logarithm of a quantity having dimensions (for example, units such as watts, pounds or feet). To get around this objection, one can consider only looking at ratios of two quantities having the same units, so that the dimensions cancel out. Thus, to calculate the value of 2 W in dB, consider that the 2 W is first divided by 1 W, so the logarithm is only taken of the dimensionless number 2. After doing this, however, one refers to the value as dBW to remember that the reference in this case was 1 W. Likewise, one can calculate the value of a frequency F Hz in decibels, by considering that it is a ratio with respect to a frequency of 1 Hz. Thus F, in decibels, is $10\log_{10}\{F\}$ dBHz (in decibels).

Why do communications engineers go to all this trouble? One reason is a property of logarithms that says that the logarithm of a product of two numbers is equal to the sum of their logarithms, and the logarithm of the ratio of two numbers is the difference of their logarithms. Because it is easier to add and subtract than to multiply and divide, using decibels makes the bookkeeping part of a link budget easier to handle.

Now, following this digression, we can return to the subject at hand. Recognizing that the product P_tG_t is the EIRP of the transmitter, Eq. (2.12) can be written in decibel form normally used in link budgets simply as

$$P_r = EIRP + G_r - L \quad \text{(all in dB)} \tag{2.13}$$

This is the most basic form of a link budget.

In practice, a real link does have other, very real losses that must also be considered. While free space does not absorb radiation, the atmosphere, particularly if it has a high moisture content or rain, does absorb radiation. Thus, both on a satellite uplink and downlink, there will be attenuation that must be accounted for in a link budget. Within the Earth station, there will be losses in the wave guide and feed between the high-power amplifier and the transmit antenna. If the antenna is slightly mispointed, so that the satellite is not exactly in the boresight direction, there will also be a reduction in the power reaching the satellite. To be exact, losses also occur if there are small errors in pointing the satellite's receive and transmit antennas. However because these are generally much smaller antennas than for the uplink antenna of an Earth station (and hence have much wider beams), these can usually be neglected. Losses also occur in the satellite repeater itself, between its power amplifier and its transmit antenna. All the losses previously mentioned (unless they are small enough to be ignored) should be included in a rigorous link budget.

E. System Noise

Regardless of how pure the modulated signal is when it leaves a transmitter, when it arrives at the receiving antenna it will be contaminated with noise. Although there are many ways to generate electromagnetic radiation, all matter produces a certain level and frequency of broadband radiation, simply by virtue of its absolute temperature. The hotter a body is, the higher the intensity of the radiation is that it emits. All electronic devices, therefore, be they active or passive, generate such thermal noise with a power level at the frequencies of interest in satellite communications, which is given by

$$P_n = kT_nB \tag{2.14}$$

where P_n is the noise power in a bandwidth B Hz wide, T_n is the absolute temperature of the body (in deg Kelvin) and k is the Boltzmann constant (1.3803×10^{-23} W/K/Hz or, in dB, -228.6 dBW/K/Hz) (Ref. 12).

F. Figures of Merit

When an antenna receives a signal from an Earth station (with a power level generally denoted as C, for "carrier") and passes it on to a low-noise amplifier, what is received at the LNA is both the received signal power, amplified by the gain of the antenna, and a certain level of thermal noise power, denoted as N. The ratio of these two quantities is an important figure of merit for the overall system.

Equations (2.12) and (2.14) give, respectively, the received signal power ($C = P_r$) and the noise level ($N = P_n$) of the antenna. We can therefore write

$$\frac{C}{N} = \frac{P_t G_t G_r}{\{4\pi R/\lambda\}^2} \left\{\frac{1}{kTsB}\right\}$$
$$= \left\{\frac{EIRP}{(4\pi R/\lambda)^2}\right\}\left\{\frac{G_r}{Ts}\right\}\left\{\frac{1}{kB}\right\} \tag{2.15}$$

where Ts is the equivalent system noise temperature of the antenna, the feed and waveguides leading to the LNA. In terms of decibels, this can be written as

$$\frac{C}{N} = EIRP - L - k - B + \left(\frac{G}{T}\right) \quad \text{(all in dB)} \tag{2.16}$$

Alternatively, rather than considering the total noise power in bandwidth B, one can instead look at the noise power spectral density, N_0, defined as the total noise power in bandwidth B, divided by the bandwidth (therefore with dimensions of W/Hz). In terms of this parameter, Eq. (2.16) can be written as

$$\frac{C}{N_0} = EIRP - L - k + \left(\frac{G}{T}\right) \tag{2.17}$$

Digital transmission waveforms convey signals as symbols, with energy transmitted per symbol, E_s. If we denote the rate at which symbols are being sent as R_s, we can evaluate the signal to noise ratio per symbol as

$$\frac{E_s}{N_0} = \frac{(C/N_0)}{R_s} \tag{2.18}$$

This can be written in decibel form as

$$\frac{E_s}{N_0} = EIRP - L - k + \left(\frac{G}{T}\right) - R_s \tag{2.19}$$

Generally, it is the transmission of bits that is of interest. The energy to transmit bits is related to that to transmit symbols by

$$\frac{E_b}{N_0} = \frac{\{E_s/N_0\}}{\log_2 M} \tag{2.20}$$

where M is the number of bits conveyed by each symbol.

Depending upon the circumstances, Eqs. (2.16), (2.17), and (2.19) are referred to as the system link equations. In either case, we see that a measure of system quality (C/N, C/N_0, or E_s/N_0) is expressed in terms of a figure of merit for the transmitter (EIRP), a figure of merit for the receiver (G/T), and the parameters of the environment between them.

For digital transmissions, a parameter called the bit error rate (BER) is often used as a figure of merit. This is a statistical parameter that expresses the probability of how often a transmitted bit in a digital bit stream is received incorrectly because of errors in transmission, noise, or interference. Bit error rate is a rather complicated function of the type of modulation and coding used for a transmission, as well as the value of system E_s/N_0.

In specifying how well a digital communications satellite system must perform, the BER is often specified. The system designer then chooses design values of EIRP and G/T, together with a defined type of modulation and coding parameters, to be able to achieve this value of BER. In practice, the designer chooses such values and parameters to exceed the specified BER. The extra power manifests itself as a link margin in the system link equations.

VI. Summary and Conclusions

This chapter has attempted to acquaint the reader with a sufficient understanding of basic communication satellite fundamentals to ensure that the concepts put forward in succeeding chapters will also be understandable. Where possible, this has been done in a qualitative fashion, avoiding mathematical rigor. When discussing the topics of antennas and link budgets, however, I have had to deviate from this goal somewhat, to more fully explain the concepts involved. In these cases, I have tried to explain the basis for the equations presented so that they will be easier to understand. In this, I hope I have succeeded, but only the readers will be able to tell.

If any reader wishes to delve further into the topics addressed herein, the list of references given hereafter would be an excellent place to start.

References

[1]*University Desk Encyclopedia*, Elsevier, New York, 1977.

[2]Agrawal, B. N., *Design of Geosynchronous Spacecraft*, Prentice-Hall, Upper Saddle River, NJ, 1986.

[3]Clarke, A. C., “Extra Terrestrial Relays,” *Wireless World*, Oct. 1945.

[4]Berlin, P., *The Geostationary Applications Satellite*, Cambridge Univ. Press, New York, 1988.

[5]Pattan, B., “Delta velocities and propulsion requirements for east-west and north-south station keeping for spacecraft in GSO,” *International Journal of Space Communications*, Vol. 17, No. 4, 2001.

[6]Fortescue, P., and Stark, J., *Spacecraft Systems Engineering*, Wiley, New York, 1995.

[7]Long, M., *The Digital Satellite TV Handbook*, Newnes, 1999.

[8]Williamson, M., *The Communications Satellite*, IOP Publishing, Ltd, 1990.

[9]Iida, T. (ed), *Satellite Communications—System and its Design Technology*, IOS Press, 2000.

[10]Course material from “SpaceTech Masters Degree in Space Systems Engineering,” Technical Univ. of Delft, The Netherlands, 2002.

[11]Richharia, M., *Satellite Communications Systems*, 2nd ed., McGraw-Hill/Macmillan Press, New York, 1999, pp. 97–98.

[12]Maral, G., and Bousquet, M., *Satellite Communications Systems*, 3rd ed., Wiley, New York, 1998.

Questions for Discussion

1) **What advantages do digital forms of modulation have over analog? Given these advantages, why weren't television and radio broadcasts modulated digitally from the beginning?**

2) Newspapers sometimes write about military surveillance satellites being relocated in orbit to "hover over" an area of military concern. **Explain why, in general, this cannot be done, and for which cases it might indeed be feasible.**

3) **Would all the methods of compressing television signals described in this chapter work equally well with sound signals? If not, which methods could be used?**

4) **What is the EIRP of a 30-GHz transmitter having an RF output power to the antenna of 20 W, when the antenna has a diameter of 1.0 m, and an efficiency of 60%?**

5) **Calculate your last salary increase in dBs.**

Chapter 3

Internet and New Broadband Satellite Capabilities

Naoto Kadowaki*
Communications Research Laboratory, Tokyo, Japan

I. Introduction

USAGE of the Internet has been rapidly growing since the early 1990s. Simultaneously, new types of services have been developed, including services such as streaming video that requires much more bandwidth than traditional services such as e-mail or simple file exchange. Because of these circumstances, the capacity of networks has been increasing rapidly. This trend has been led by fiber optics technology coming into wide usage in terrestrial networks. On the other hand, satellite communications systems have been used in various kinds of applications, and satellite usage for Internet access has shown significant increases in recent years. The importance of satellite communications systems in this area can be summarized as

1) Rapid introduction of broadband Internet services where terrestrial communication infrastructure is poor.
2) Backing up of terrestrial communication infrastructure destroyed by disasters to maintain the lifelines of emergency communications.
3) Providing opportunities for new services that use the unique advantages of satellite communications systems such as their multicast capability.
4) Bypassing highly complicated terrestrial interconnection of networks to provide direct multipoint connections distributed over wide areas.

This chapter describes the technologies necessary to realize broadband satellite networks and satellite Internet applications. A few technical demonstrations are also introduced.

---

*Group Leader of Broadband Satellite Network Group, Wireless Communications Division.

II. Key Technologies for Broadband Satellite Communications

A. Wide Bandwidth Transponder

To realize broadband satellite communications systems, wide frequency bandwidth is required. This is, in general, only available at Ku-band or higher-frequency bands, which therefore should be used for broadband applications.

Ku-band is already widely used for fixed satellite services (FSS). It is thus difficult in most areas to introduce new services that require wide frequency resources. Most of the Ku-band satellite transponders that are commercially available have only 27-MHz or 36-MHz bandwidths, although two transponders can be combined as a wider-bandwidth transponder, so that a virtual 72-MHz transponder can be configured from two 36-MHz transponders. Intelsat satellites, such as the Intelsat-VIII series, have much wider bandwidth transponders with up to 108 MHz or 120 MHz bandwidth in Ku-band. Broadband satellite communications systems can thus be realized directly with these wide-bandwidth transponders. Figure 3.1 shows the frequency allocation of Ku-band transponders of Intelsat-VIII as an example.

Ka-band is more likely to be used for broadband applications because of its wider allocated bandwidth, and because it is not as widely used for current services compared with the Ku-band. Japan launched the first Ka-band communications satellite, Sakura, in 1977, which had six 200-MHz bandwidth transponders in the Ka-band. The current commercial Ka-band satellite in Japan is the N-STAR, launched in 1995, which has a bandwidth of 200 MHz for its domestic multibeam transponders and 100 MHz for its domestic shaped-beam transponder. The Communications Research Laboratory (CRL) carried out broadband satellite communications experiments with the N-STAR multibeam transponders, and these experiences will be described later in this chapter.

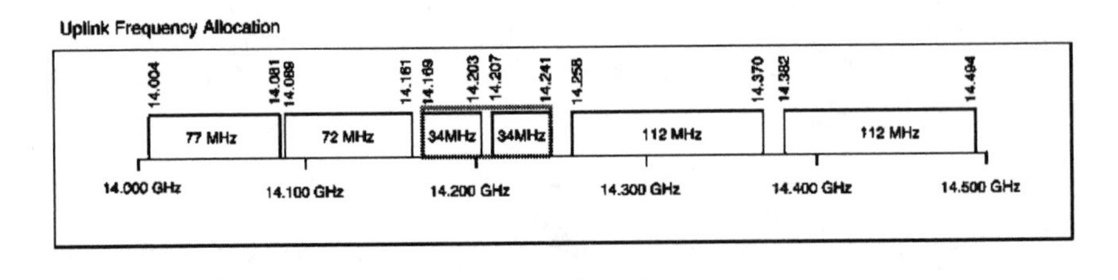

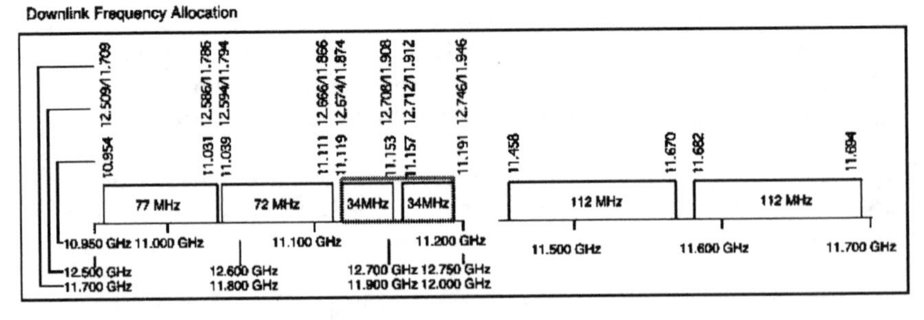

Fig. 3.1 Frequency allocation of Intelsat-VIII.

NASA demonstrated a much higher data rate satellite communication system with its Advanced Communications Technology Satellite (ACTS),[1] which was launched in 1993. The ACTS was equipped with a Ka-band transponder whose bandwidth was 900 MHz and could carry both 622 Mbps (STM-4 or OC-12) information rates and 155 Mbps (STM-1 or OC-3c).

To promote high-speed Internet by broadband satellite communications systems, the National Space Development Agency of Japan (NASDA) and CRL are jointly developing an experimental communications satellite, wideband internetworking engineering test and demonstration satellite (WINDS), which is scheduled to be launched in late 2005 (Ref. 2). The WINDS will carry Ka-band transponders that use the 27.5 GHz to 28.6 GHz band for uplink and 17.7 GHz to 18.8 GHz for downlink. The WINDS will have the capability to carry a 1.2 Gbps (STM-8 or OC-24) information rate with bent-pipe connection through a 1.1-GHz bandwidth transponder. The external view of WINDS is shown in Fig. 3.2.

B. Multibeam Antennas

To realize high data rate transmissions, high-gain antennas are required. This means that a satellite's onboard antenna must generate a narrow beam width. In such a case, a multispot-beam antenna is required to provide both high gain and wide coverage simultaneously.

There are two types of multispot-beam antennas. One is a fixed multispot-beam antenna (FMBA) and the other is a steerable multispot-beam antenna (SMBA). An FMBA can be configured by a parabolic reflector with a multihorn feed, where each horn covers a fixed area on the Earth. Most current satellites that have multibeam antennas use this type of antenna. Its configuration is relatively simple, but if a satellite must have full Earth coverage, a large number of feed horns and several reflectors are required. Therefore, this type seems to be most suitable for domestic or regional satellites.

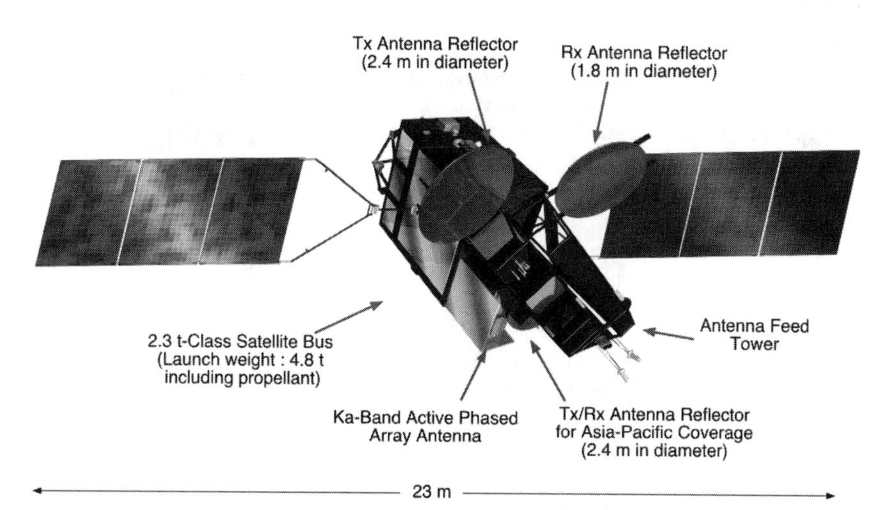

Fig. 3.2 External view of WINDS.

SMBAs can be divided into two types—mechanically steerable antennas and electronically steerable antennas. Mechanically steerable antennas can be realized by adding a gimbal mechanism to an ordinary dish antenna. This type can be manufactured easily, but beam steering capability is limited by the driving speed of the mechanism. Electrically steerable antennas require phased-array feeds. To achieve higher gain, these may be used with reflectors, but the direct radiating types are also used where their gain is sufficient. Electronically steerable antennas can scan beams very quickly with no mechanical movement required. In terms of the beam scanning angle, direct radiating types have the widest scanning capability, the imaging reflector type is the next, and the single reflector type is the third.

The Japanese experimental satellite, WINDS, will be the first satellite equipped with a Ka-band direct radiating type of active phased-array antenna. If the functions and performance of this antenna are verified, direct radiating antennas will be a new option of onboard antenna in the future.

C. Satellite Onboard Switching and Routing

As described later in this chapter, an onboard switching and routing function provides flexibility in satellite networking. Therefore, several satellite systems for which plans were announced in the mid 1990s use onboard switching and routing capability.

NASA's ACTS demonstrated the functions and performance of onboard switching. Following ACTS, planned commercial satellite systems such as Spaceway, Astrolink, and Teledesic used onboard switching and routing technology. For example, Astrolink adopted Asynchronous Transfer Mode (ATM) switching technology, which is widely used in high data rate terrestrial networks.[3]

Japan's WINDS also has an onboard ATM switch subsystem.[4] The major specifications of the onboard ATM switch for the WINDS are shown in Table 3.1.

Table 3.1 Major specifications of the onboard ATM switch for WINDS

Demodulator	Type	Digital signal processing
(D-DEM)	Data rate	1.5 Mbps (by freq.-demux), 6.1, 24.0, 51.84 Mbps/unit 155.52 Mbps (with 3 units)
	FEC	RS(255, 223)
	Number of units	9
ATM switch	Data rate	155.52 Mbps/port
(ATMS)	Throughput	2.4 Gbps
	Number of ports	3 × Input, 3 × Output
	Number of SW	1 + 1(back up)
Modulator	Type	Analog/DSP Hybrid
(MOD)	Data rate	155.52 Mbps/unit
	Number of units	3

The throughput of the switch is 2.4 Gbps, and the switching architecture is designed to be modular-based, so that the throughput can be multiplied by adding switching modules. Because WINDS is an experimental satellite, the switch has only three input/output lines. An onboard demodulator and modulator are additional components to complete a full onboard switch and router. WINDS will be equipped with quadrature phase shift keying (QPSK) demodulators and modulators. The demodulators are configured mainly by field programmable gate array (FPGA) devices, and digital signal processing technology is used. The downconverted input signal is led to an orthogonal detector, then the I-Q divided signals are analog–digital converted. These digital signal streams are demodulated by a digital signal processing in FPGA devices. This demodulator can demodulate up to a 51.84 Mbps information rate signal at the maximum. For providing multichannel demodulation capability, a frequency demultiplexer that is configured by subfiltering and fast Fourier transform (FFT) processing can be inserted before the demodulation process. As the result, each demodulator can demodulate 14 channels × 1.5 Mbps and single channels at 6 Mbps, 24 Mbps, and 51.84 Mbps. The onboard modulator is also designed with digital signal processing technology to generate modulated signals, with an information transmission rate of up to 155.52 Mbps.

The onboard ATM switch can provide a layer-2 routing function. However, Internet users exchange data using IP (Internet protocol) as a layer-3 protocol and TCP (transport control protocol) as a level-4 protocol. Recent trends in the Internet community show that the layer-2 protocol is not always necessary, and it is preferable to avoid it and its attendant overhead. An example of this trend is the IP over Synchronous Optical Network (SONET), which is an implementation of IP directly over the physical layer protocol, SONET, skipping the layer-2 protocol. From these points of view, using the ATM protocol for onboard switching and routing is perhaps somewhat controversial. An onboard IP router is essential to realize an equivalent protocol stack in broadband satellite networks; this will probably become a major research item in the future.

D. Ground Segment Technology

To realize broadband satellite networks, the use of Ku or higher frequency bands is necessary. Technical advancements in Ku-band Earth stations provide a near-term solution for broadband satellite networking, and those in Ka-band will provide essential solutions for the very high data rate transmissions, (for example, up to 1.2 Gbps), which are required to carry huge amounts of data to and from many users. In any case, Earth stations should have high equivalent isotropically radiated power (EIRP) and gain to temperature ratio (G/T) over wide bandwidths. Modulation and coding schemes are also very important because a satellite communications link is limited by both power and frequency resources. Accordingly, the following are items to be implemented:

1) High-power transmitter: for example, a few hundred watts in Ku/Ka-band.
2) Low noise amplifier for receivers: for example, NF (Noise Figure) of 3 dB or less over 2 GHz bandwidth.

3) Flatness of amplitude and phase characteristics in frequency upconverters and downconverters.
4) Very high speed modulation: for example, multibit/Hz modulation schemes.
5) Very high speed and efficient coding hardware: for example, a chip set for Turbo code at a few Gbps.

To realize user-friendly broadband Earth terminals, compact and lightweight implementation is required. The following are examples of the technologies needed:

1) Compact antennas: for example, 45 cm diameter equivalent lightweight antennas in Ka-band.
2) Simple or automatic but inexpensive satellite tracking function.
3) Less-expensive transmitters: for example, Solid State Power Amplifier (SSPA) in Ka-band or higher frequency bands.
4) Common user terminal interface: for example, wireless LAN, or Gigabit Ethernet.

III. Networking Architecture with Satellites

A. Networking with a Bent-Pipe Transponder

In the case that a bent-pipe transponder with IF switching function is used, the IP routing function can be managed by ground-based network control stations and user terminals. This networking concept is shown in Fig. 3.3. Because the connection is controlled by the NCS, establishing a connection experiences a long delay, which could degrade the efficiency of capacity use. In addition, all user terminals should be connected to a router in a satellite network directly, and all routers should be connected to one or more routers. In this case, the gateway

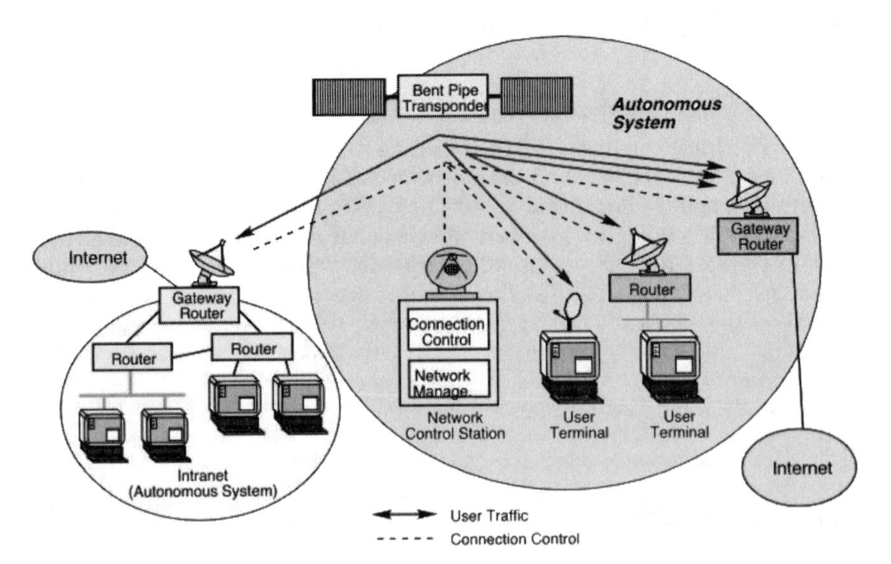

Fig. 3.3 Networking concept with bent-pipe transponder.

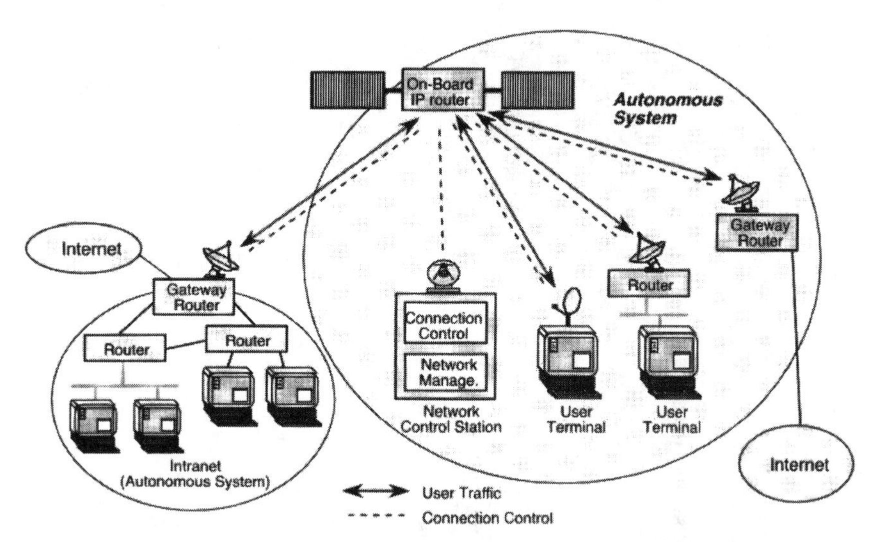

Fig. 3.4 **Networking concept with onboard routing transponder.**

stations, which have routers, should have the capability to handle many physical satellite links to communicate with users in corresponding subnets. In addition, at least one border gateway station should be in the satellite network to communicate with autonomous systems outside of this satellite network.

B. Networking with a Regenerative Transponder

An onboard IP routing function is one of the most effective schemes for highly efficient internetworking. This networking concept is shown in Fig. 3.4. All terminals and routers have one physical link to the satellite link because the satellite itself behaves as a router. Though the connection control link should be prepared independently, the network linking can be much more simple than that in Fig. 3.3. The ground-based border gateway is not always necessary because the onboard router can behave as a border gateway.

Considering the onboard processing technology currently available, developing an onboard, high-throughout IP router or layer-3 switch involves several difficulties. One is that such a function should be implemented in combination with a high-speed processor and large-scale software, but it is hard to devise a high-speed processor for onboard applications. Another is that the layer-3 switching technology is changing rapidly, and this means that any onboard subsystem built with the current technology may soon be obsolete. Regarding layer-2 switching technology, an onboard ATM switch has been developed in Japan's Gigabit Satellite Project[5] and the Gen*Star Project by TRW.[6] It seems feasible that the technology may be ready for application around 2005.

IV. Protocols and Standards

Satellite communications links have some different characteristics from terrestrial communication links, such as fiber-optic cables. For example, in a

Table 3.2 Standardization activities relating to satellite-based Internet

Organization	Established standards or activities	Remarks
ETSI	(1) TR 101 374-1: "Satellite Earth Station and Systems (SES); Broadband satellite multimedia; Part 1: Survey of standardization objectives," July 1998 (2) TR 101 374-2: "Satellite Earth Station and Systems (SES); Broadband satellite multimedia; Part 2: Survey of standardization scenario," March 2000 (3) EN 301 459, v.1.2.1 (2000-08): "Harmonized EN for Satellite Interactive Terminals (SIT) and Satellite User Terminals (SUT) transmitting towards satellites in geostationary orbit in the 29.5 to 30.0 GHz frequency bands covering essential requirements under article 3.2 of the R&TTE Directive"	Based on EN 301 358 (1999) and EN 301 359 (1999)
	<DVB related> (1) EN 300 421 "Digital broadcasting systems for television, sound and data services; Framing structure, channel coding and modulation for 11/12 GHz satellite services" (2) TR 101 790 V.1.2.2 (2000-12): "Digital Video Broadcasting (DVB); Interaction Channel for Satellite Distribution Systems; Guidelines for the use of EN 301 790"	DVB-RCS001 ver14 (03 April 2000) was published.
TIA	"TR 34: Satellite Equipment and Systems" proceeds study of satellite networking protocols. TR34.1: "Communications and Interoperability" TIA/EIA/IS-787: "Common Air Interface for Satellite Interface (CASI) Interoperability Specification" Following items are studied: – Internet Protocol over Satellite – ATM Traffic management and congestion control – ATM Speech – ATM Multicast over Satellite – Common Air Interface: ATM via Satellite – Common Air Interface: Dual Mode GSM-Compatible GEO Mobile System – Ka-Band Satellite Systems – IP QoS Architectures for Satellite Networks	

(continued)

Table 3.2 Standardization activities relating to satellite-based Internet (continued)

Organization	Established standards or activities	Remarks
	TR 34.2: "Spectrum and Orbit Utilization" Contribution to FCC and ITU.	
IETF	TCPSAT-WG established RFSs for use of satellite-based Internet. (1) RFC2488: "Enhancing TCP over satellite Channels using Standard Mechanisms" (2) RFC2760: "Ongoing TCP Research Related to Satellites" Several working groups study protocols that can be utilized for satellite networks. (1) UniDirectional Link Routing (UDLR) WG (2) Performance Implications of Link Conditions (PILC) WG (3) Audio Video Transport WG (4) Reliable Multicast Transport WG	[Related RFC] RFC1323, RFC2018, etc.
ITU	ITU-T studies the following items: (1) Y.1540 (I.380): "Internet Protocol Data Communication Service - IP Packet Transfer and Availability Performance Parameters" (2) Y.1541: "Network performance objectives for IP-based service" (3) Y.1231: "IP Access Network Architecture" ITU-R started studies regarding the following items: (1) "Performance for B-ISDN ATM via Satellite" (2) "Availability for ATM via Satellite" (3) "Satellite Link Performance for Transmission of IP" "Intersector Coordination Group on Satellite Matters (ICGSAT)" studies the following items. (1) "IP over Satellite Matters" (2) "Convergence"	

satellite link, burst errors often occur when the link quality is degraded. On the contrary, the error rate is much smaller in fiber-optic cables and random error is dominant if errors occur. In addition, a geostationary (GEO) satellite link has a one-way, 250-ms delay. This delay affects communication efficiency when a handshake-type protocol is used. Therefore, each layer of a protocol stack has to be designed to avoid the influence of such errors and delay.

Standardization of protocols is another key for diffusion of satellite-based communication services. Table 3.2 shows the standardization activities for broadband satellite networking protocols by many organizations in the world. The European Telecommunications Standards Institute (ETSI) made great efforts to use the digital video broadcasting (DVB) infrastructure for satellite Internet and to establish DVB-S/DVB-RCS protocol standards. The Telecommunication Industry Association (TIA) in the United States standardized an ATM transmission scheme via satellite. The Internet Engineering Task Force (IETF) established the transmission control protocol over satellite (TCPSAT) working group, and investigated how to apply TCP to satellite links. One of the purposes of the TCPSAT-WG was to promote mutual understanding between the Internet community and the satellite communication community. NASA Glenn Research Center contributed greatly to the TCPSAT-WG activities. The TCPSAT-WG finished its role in 1999, but related items have since been investigated. The International Telecommunication Unit (ITU) started to study satellite-based Internet standards, and both the ITU-T and ITU-R initiated study groups related to satellite Internet protocols. The following sections discuss considerations of protocol specifications for satellite-based Internet.

A. Layer 1

Layer 1 (the physical layer) specifies mainly satellite access schemes, modulation schemes, error correction codes, and other techniques necessary for providing transmission capability. We can select modulation and coding schemes from several combinations that are typically used. On the other hand, the characteristics of Internet applications should be considered when a satellite access scheme is selected. In the nature of Internet applications, users can typically access any kind of server, such as mail servers or web servers. Servers usually send large amounts of data to users. In addition, servers often send the same information to multiple users. In this case, a time division multiplex (TDM) scheme is appropriate for the link from the server to users (called the forward link). For the links from users to server (the return link), frequency division multiple access (FDMA) may be the simplest choice. To maximize the capacity use, time division multiple access (TDMA) is a candidate. For example, DVB-S uses TDM as a forward link, and DVB-RCS uses MF-TDMA as return links (see Chapter 5 for further information).

B. Layer 2

Layer 2 (the link layer) specifies a procedure of logical connection between Earth stations (nodes). This layer includes procedures for the establishment of a logical link, data transfer, and termination of the logical link. Sometimes, it includes error recovery procedures between nodes known as automatic repeat request (ARQ). High-level data link control (HDLC) or synchronous data link control (SDLC) are traditional examples of this layer of protocols, and ATM has recently become the most popular layer-2 protocol for high data rate networks.

C. Layer 3

Layer 3 (the network layer) specifies procedures to provide terminal-to-terminal connections. Considering satellite-based Internet, any IP routing function is in this layer. Terminals are not necessarily directly connected to an Earth station, but can be located within a local area network (LAN) or wide area network (WAN) connected to an Earth station. After a layer-2 link between Earth stations is established, IP routing through satellite links is available with one or more routers. Multiple autonomous systems (AS) can be implemented with a satellite network if one or more border gateways are implemented, and systems outside of the satellite network can be accessible through the border gateway.

D. Layer 4

In the Internet connection, layer 4 (the transport layer) is specified as TCP or user-defined protocol (UDP). TCP was designed to improve the reliability of linking, because IP cannot guarantee the reception of sent packets. To verify the reception at a destination terminal, TCP requires window-based acknowledgment from the receiving terminal. Most typical Internet terminals have 64-kB windows for waiting acknowledgment, and if a sending terminal does not receive acknowledgment from the receiving terminal after it sends 64-kB data, it stops sending and waits until it receives the acknowledgment.

Considering that a GEO satellite link has 500-ms, round-trip delay, the maximum transmission efficiency (which is called a throughput) is 128 kB/s when the window size is 64 kB. To break this barrier, the window size must be enlarged. In addition, some other modifications were made by the TCPSAT working group in the IETF, and an improved protocol for a "long-fat-pipe" (which means high data rate and long delay link) was released as RFC-2488,[7] the so-called TCP extension. RFC-1323 specifies window scaling and time stamp options, and RFC-2018 specifies the selective acknowledgment scheme.

Even with these extensions, the TCP has other problems to limit throughput. It uses a slow-start procedure. The slow-start procedure behaves as follows. The sending terminal sends only one packet initially, and if an acknowledgment is verified, then the sending terminal sends two packets. The sending terminal iterates this procedure, doubling the number of packets sent each time as long as an acknowledgment is sent back. When a certain "time-out" limit is received without the sending terminal having received an acknowledgment, it then halves the number of packets sent per transmission and tries again. If it then receives an acknowledgment, the sending terminal gradually increases the number of packets sent each time, one by one, until a maximum transmission rate is found, above which a lack of reception of acknowledgments is again experienced. This procedure is referred to as the congestion control procedure. When the round-trip delay is relatively large, as it is in a GEO satellite link, it takes quite a long time to achieve an appropriately high data rate, because it takes one-half second to get an acknowledgment from the receiving terminal every time.

On the other hand, UDP is opened for a user's definition of usage. With it, users can specify any kind of procedure for their own purposes. That means proprietary protocols can be implemented as a usage of UDP instead of TCP.

In addition to TCP and UDP, other transport layer protocols have been proposed. One example of such protocol is the express transport protocol (XTP). XTP uses a large window size and avoids slow start/congestion control procedures by registering the available maximum transmission rate and expected latency between nodes. A method of protocol conversion at Earth stations between TCP and this kind of protocol for long-fat-pipes is called a performance enhancing proxy (PEP) and is recognized as one of the most important study items for Internet implementation in very high-speed networks.

V. Technical and Application Demonstrations

A. ATM Transmission Experiments with N-STAR

CRL conducted ATM transmission experiments over high data rate (HDR) satellite links using an operational satellite, N-STAR-a, between 1996 and 1998.[8] To realize satellite-based HDR information networks, there are some technical issues to be evaluated. The required link quality, quality of services and networking protocols in multimedia networks are very different from those in ordinary telephone networks. In addition, the ATM and synchronous digital hierarchy (SDH) protocols were originally developed for fiber-optic networks. Therefore, it was very important to evaluate whether these protocols would work properly via satellite links that have different error characteristics than fiber-optic links. ATM/SDH transmission performance such as the STM-1 synchronization performance, cell loss, cell header error, data bit error, and others, in given bit error rate (BER) situations, were evaluated in this experiment.

1. *Experimental System Configuration*

The concept of the experimental system is shown in Fig. 3.5. Two Earth stations were set up at CRL's Kashima Space Research Center (KSRC) and its Seika Communications Research Center (SCRC). These stations were connected by a satellite link provided by N-STAR's Ka-band multibeam transponder with an information transmission rate of 155.52 Mbps. The transponder covers Japan with eight uplink spot beams and three downlink spot beams. The transponder bandwidth is 200 MHz. The EIRP and G/T are 51 dBW and 14 dB/K, respectively. Left-hand circular polarization and right-hand circular polarization are used for uplink and downlink, respectively.

The HDR Earth station was configured using a 5-m-diameter antenna, a high-power amplifier (HPA), frequency upconverters and downconverters, a low-noise amplifier (LNA), and a QPSK modem (modulator and demodulator). The HPA used a coupled cavity type TWT with an output power of 350 W. The LNA used a monolithic integrated circuit (MIC) type configuration and achieved an NF of 2.5 dB. As a result, the EIRP and G/T were 74.0 dBW and 31.1 dB/K, respectively. The QPSK modem could carry the information at 155.52 Mbps, which is

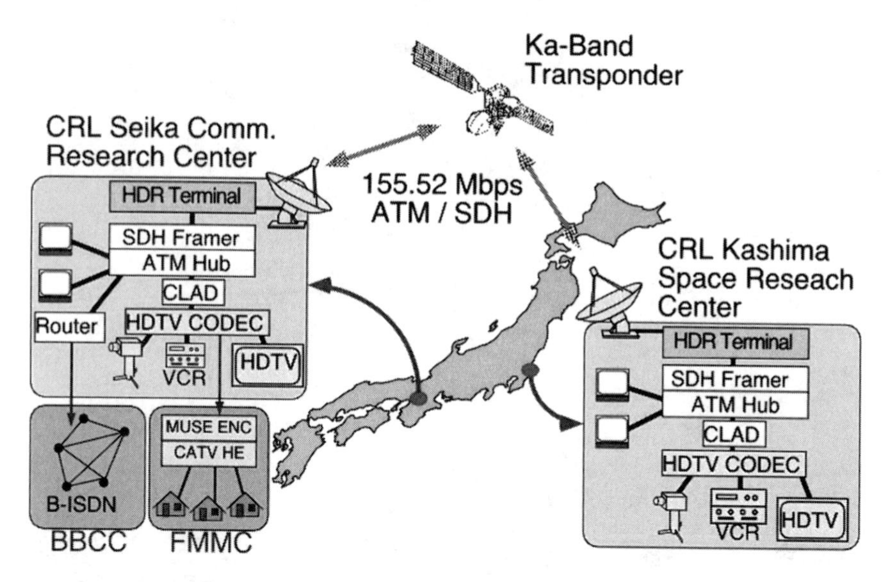

Fig. 3.5 Experimental system of high data rate ATM networking using N-STAR.

compatible with the STM-1 specification. The transmitted information was coded by a Reed–Solomon (255, 223) code and mapped into the Consultative Committee for Space Data Systems (CCSDS) format with 8-bit symbol interleaving with a depth from 1 to 5. Then, the transmission rate over the satellite link was 189.54 Mbps.

All data transmitted through the satellite link was formatted in ATM cells and multiplexed into an STM-1 frame. The configuration of the ATM/SDH facility is shown in Fig. 3.6. The ATM hub that was set up in each site has four user network interface ports to connect workstations and one network node interface port to connect to another hub via the satellite link. The transmission line interface between the ATM hub and satellite modem was STM-1, specified by SDH protocol. Workstations connected to the ATM-hub had built-in cell level assembly and disassembly (CLAD) cards that used an ATM adaptation layer type 5 protocol and 140 Mbps transport asynchronous transmitter/receiver interface. This small-scale ATM-LAN was used for ATM traffic generation and experimental data gathering, and also served as a multimedia communications terminal. In the SCRC, a router that had an ATM interface was installed to connect to an experimental broadband integrated services digital network. To carry a compressed digital HDTV signal, a dedicated CLAD based on AAL type 1 protocol was also installed along with the ATM hub.

In both the KSRC and the SCRC there are HDTV studios equipped with HDTV cameras, video tape recorders, an editing console, and monitors. The raw HDTV signals are digitized and compressed into 131.072 Mbps by a video CODEC (coder and decoder) that uses discrete cosine transform and motion compensation techniques. Using these facilities, HDTV programs could be transferred between

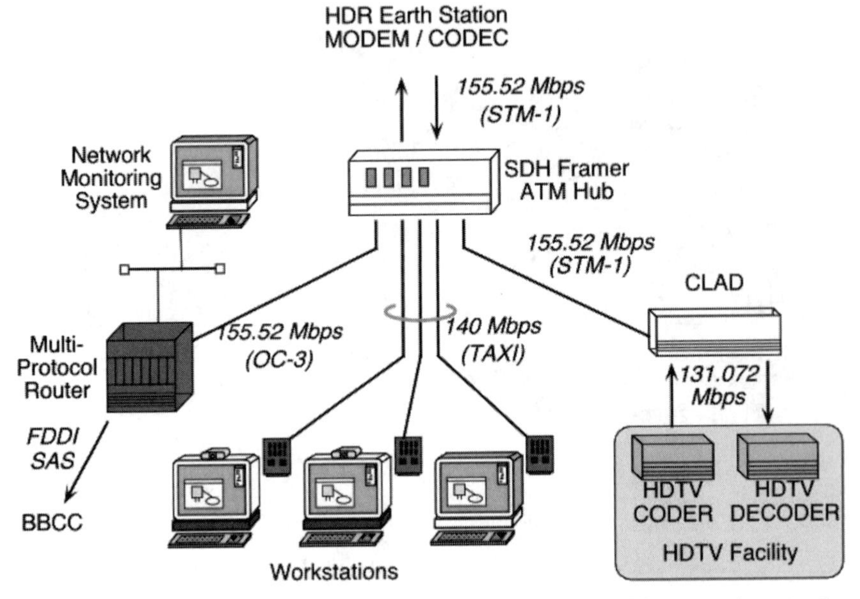

Fig. 3.6 ATM/SDH facility for high data rate ATM networking experiment using N-STAR.

KSRC and SCRC. HDTV signals could be also converted to analog multiple sub-nyquist sampling and encoding signals in the SCRC and transmitted to 300 homes in the neighborhood of SCRC via a fiber-optic CATV pilot network.

2. *Experimental Result*

The BER under certain C/N_0 situations and cell loss ratio (CLR) under certain BER situations were measured to evaluate ATM transmission performance over the HDR satellite communications link. The depth of the symbol interleaver (ITL) and permitted number of bits in the CCSDS frame header were changed as parameters during the measurements.

The BER characteristics were measured by reducing transmission power, with the results shown in Fig. 3.7. From this result, it could be seen that the BER performance in the satellite loop-back mode was greatly degraded compared to the performance in the IF loop-back situation. One of the possible reasons for this degradation may be the nonlinearity of the satellite transponder, but we have to consider the method of noise power estimation in nonlinear systems. We continue to investigate it. In addition, BER changes dramatically with very small changes of C/N_0 because forward error correction (FEC) works very well to correct almost all errors when C/N_0 is better than a certain level, but is not effective if C/N_0 is not better than that level.

The cell loss performance obtained is shown in Fig. 3.8. The abscissa shows the BER that is predicted using 24 bit-sequence of Bit Interleaved Parity (BIP-24)

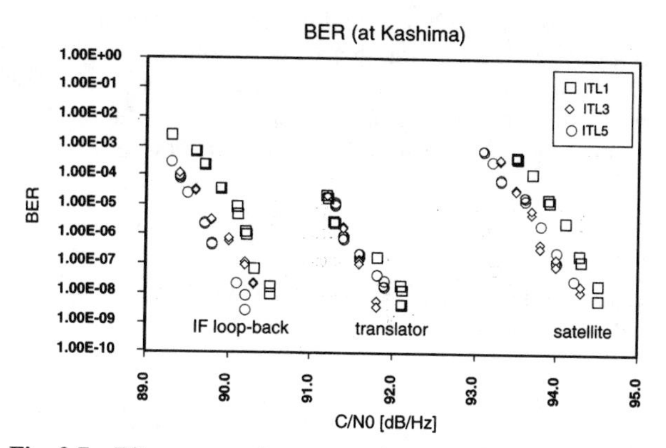

Fig. 3.7 Bit error performance of STM-1 link over N-STAR.

embedded into the STM-1 frame. Figure 3.8 shows that the depth of the interleaver does not affect the cell loss performance. The reason for this result is thought to be that the symbol interleaver does not work sufficiently to scatter burst errors, which leads to cell loss. An ATM cell is discarded when two or more bit errors are in the cell header, so a burst of bit errors causes a high cell loss rate.

The results of a simulation in random errors are also shown in Fig. 3.8. In this simulation, we assumed that rate $1/2$ (R$1/2$) convolutional coding and Viterbi decoding are used as an FEC and the bit-interleaving frame is 32,460 bits. Though the parameters assumed in this simulation are different from the actual experimental parameters, the simulation result is similar to the measurement result.

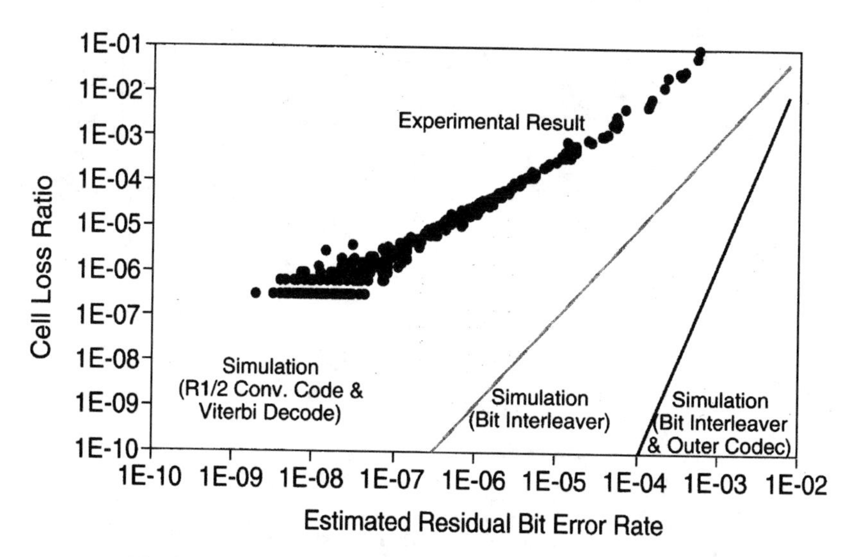

Fig. 3.8 Cell loss performance of STM-1 link over N-STAR.

Compared to the simulation result of a random error situation, the measured results are much degraded. The reason for this degradation is thought to be that the received signal contains burst errors, just like the simulated circumstances with convolutional coding/Viterbi decoding. During BER measurements, burst errors affecting approximately 20 bits occurred a number of times.

B. U.S.–Japan Trans-Pacific HDR Satellite Communication Experiments[9]

More than ten organizations in Japan and the United States [including CRL and the Jet Propulsion Laboratory (JPL)] demonstrated 45 Mbps ATM link performance with remote high definition video post-production in 1997 as part of phase 1 of the Trans-Pacific High Data Rate Satellite Communications (TP-HDR-SAT) Experiment. The TP-HDR-SAT team carried out a second phase of experiments focusing on the Internet operability over the Trans-Pacific HDR networking environment. A demonstration of remote astronomy and visible human remote access were also performed with this experimental network.

1. *Experimental Network Configuration*

Figure 3.9 shows the experimental link used for the second phase demonstration. It was provided by the combination of the NASA Research and Educational Network (NREN), ATD-NET, Pacific Bell, CA*net2, INTELSAT, N-STAR, and Japan Gigabit Network (JGN). Sapporo Medical University (SMU), CRL, Mt. Wilson Observatory, JPL, Goddard Space Flight Center (GSFC),

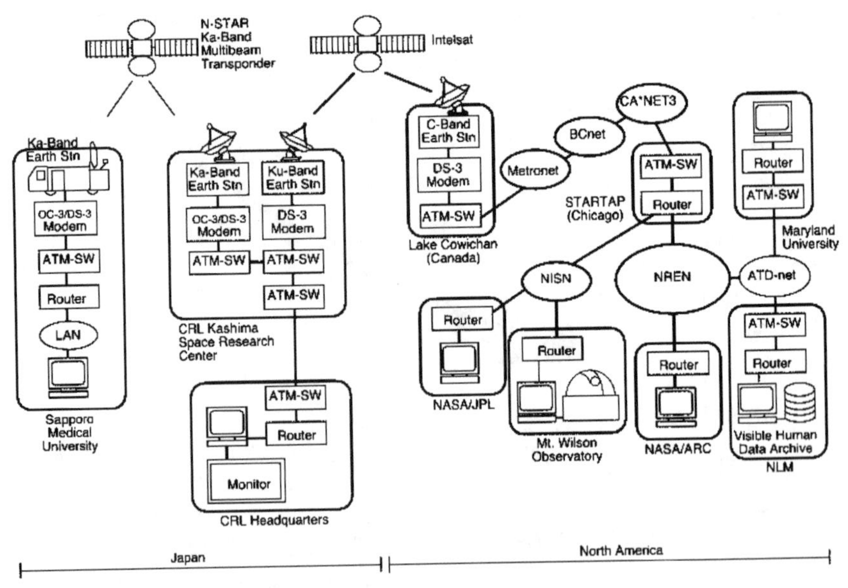

Fig. 3.9 Experimental network configuration for the Trans-Pacific HDR Satellite Communications Experiment.

National Library of Medicine (NLM), and some other sites were interconnected with this link to carry out the remote astronomy and visible human demonstrations of remote astronomy and visible human remote access. In advance of establishing the Intelsat link, the team tested the end-to-end connectivity via NREN, ATD-NET, TransPac-APAN, and IMnet/N-STAR, the latter of which is the combination of terrestrial networks without N-STAR.

2. *Result of Network Performance Tests*

The Intelsat link was established with the satellite system operation guide test, and after the satellite link at DS-3 rate was established, it was very stable and did not suffer from bit errors except when the KSRC area had heavy rain during a typhoon in July 2000.

CRL carried out engineering tests to measure ATM transmission performance via N-STAR, because the available time of Intelsat was limited. The test was done in February and March 2000 in advance of the actual Trans-Pacific demonstration, which began in late May. Figures 3.10 and 3.11 show the cell error performance over DS-3 and OC-3c satellite links, respectively.

CRL tested a QPSK modulation scheme and four sets of coding schemes with R3/4 and R7/8 convolutional codes/Viterbi decoding, and Reed–Solomon code RS(208,192) for the DS-3 satellite link. It was clear that QPSK with R3/4 convolutional code and RS(208,192) achieved the best performance. An almost error-free situation was achieved when the C/N_0 was more than 88 dB-Hz, with any set of the four coding schemes. In the actual demonstrations, 8PSK modulation with 5/6 trellis coding was used for the Intelsat

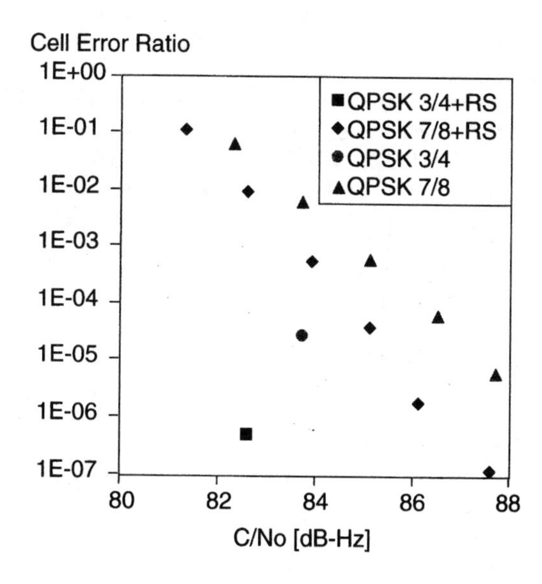

Fig. 3.10 Cell error performance of DS-3 link for the Trans-Pacific demonstration.

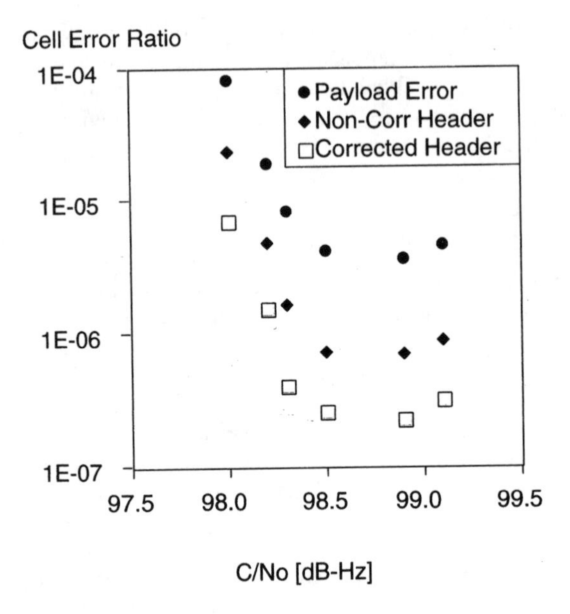

Fig. 3.11 Cell error performance of OC-3c link for the Trans-Pacific demonstration.

link, though QPSK R3/4 and RS(108,192) was used for N-STAR domestic link. Therefore, the performance of the Intelsat link seems approximately 3 dB worse than that of QPSK R7/8. We operated the Intelsat link with C/N_0 of better than 93 dB-Hz, at which point an error-free situation was achieved for the actual demonstrations.

Regarding the OC-3c satellite link, the satellite modem uses R8/9 trellis coded 8PSK and RS(255,239). No better performance could be achieved than 5×10^{-6} of cell error ratio, even if the C/N_0 was better than 98.5 dB-Hz. Though the cause of this was never made clear, it is believed that some phase noise generated in the modems caused the degradation of performance. In other experiments, much better performance was obtained with the same modems using the different route of the N-STAR's Ka-band multibeam transponder. A DS-3 link with QPSK R3/4 and RS(208,192) modulation/coding schemes was used for the N-STAR link during demonstrations. The following items were evaluated as a network performance test.

Reachability/round-trip time/jitter/packet loss. These characteristics were measured by running a series of pings from a UNIX workstation at each site to workstations at all of the other sites. The obtained data can be used to build a profile of round-trip times across different segments of the network. The delay between NLM and SMU via Intelsat was more than 1 s because the link had one more satellite link by N-STAR and Intelsat.

Hop counts/congestion/path symmetry. These characteristics were measured by running a series of "trace-routes" commands between the test workstations at

4-h intervals. These data yield some information about route stability, symmetry, and potential congestion points.

Throughput. Because this experimental network was configured by satellite links and terrestrial links, the delay-bandwidth-product (D*B) became equivalent to about 10 Mbits and 5 Mbits for links between SMU and NLM and CRL-HQ and NASA/Ames Research Center (ARC), respectively. This large D*B affects the transmission performance in the TCP or equivalent layer. TCP, UDP, and XTP were used as the layer-4 protocols, and the transmission performance of TCP and XTP were measured during the experiment.

TCP and UDP throughput was measured using the software tool "iperf." Network file system (NFS), the distributed file system protocol used by the visible human application, was run over TCP connections. As a result, the large bandwidth-delay product of the link was a major concern because the visible human application ran on an operating system whose TCP stack could not be adequately tuned to provide maximum throughput; hence, the need for the SkyX gateways. The remote astronomy application used the Andrew file system (AFS) distributed file system over UDP and was not subject to the same concerns.

The TCP measurements were taken with a window size of 64 kB, which represents the maximum window size available on machines that do not support extended window sizes. The TCP measurements between NLM and SMU over Intelsat were taken with SkyX processing active. Without SkyX, the TCP results were 200–800 kbps in each direction. Values marked with an asterisk are estimates obtained using the software tool "bing," which provides a rough estimate of throughput based on round-trip times of different-sized. Internet control message protocol (ICMP) packets. This software tool is useful in situations where an iperf-type receiving daemon is not available on the remote end.

Of particular interest is the poorly observed throughput in the TCP tests over the satellite link. The prevailing theory was that ATM cell drops, which resulted in lost TCP segments that triggered the TCP congestion control mechanisms, caused the low TCP throughput. The cell drops were further posited to be caused by non-conformance to the rate limiting parameters set on the permanent virtual circuits.

Application demonstrations. The RA demonstration was carried out connecting mainly CRL-HQ, Mt. Wilson Observatory (MWO), JPL, NASA/ARC, and the University of Maryland. High school students in Tokyo and California attended the demonstration, manipulated the telescope at MWO remotely, and were given a lecture by an astronomer at the University of Maryland. The visible human demonstration was done mainly between NLM and SMU. Medical researchers at SMU accessed the visible human archive at NLM and processed the data with their local computers. This demonstration was opened to the press at SMU. Both demonstrations were successfully done in July 2000.

VI. Conclusions

This chapter describes various technologies required to realize broadband satellite communications systems and their applications to the Internet. Very wide bandwidth transponders, spot beams antennas, and efficient modulation/coding schemes are a part of the key technologies for high data rate transmission systems. In addition, onboard switching/routing and protocols to overcome high latency and burst errors are essential to apply broadband satellite communications systems to broadband Internet. Several trials to apply satellite communications systems to broadband Internet have also been examined in this chapter. In summary, satellite communication systems have a potential to be an important part of the Internet. The means of using satellites and the best type of technology for this type of application continue to be a very hot topic in satellite communications research and development.

References

[1]Gedney, R. T., Schertler, R., and Gargione, F., *The Advanced Communications Technology Satellite*, Scitech Publishing, Inc., 2000.

[2]Kuramasu, R., Araki, T., Shimada, M., Tomita, E., Satoh, T., Kuroda, T., Yajima, M., Maeda, T., Mukai, T., Kadowaki, N., and Nakao, M., "Wideband Internetworking Engineering Test and Demonstration Satellite (WINDS) System," *Proceedings of the 20th International Communication Satellite System Conference*, May 2002; also AIAA Paper 2002-2044.

[3]Gobbi, R. L., Grant, J. D., Rosener, D. S., and Liu, M., "Astrolink: An Evolutionary Telecommunications Venture," *Proceedings of the 6th Ka-Band Utilization Conference*, Cleveland, OH, May 2000, pp. 49–54.

[4]Kadowaki, N., Yoshimura, N., Nakao, M., and Ogawa, Y., "High Speed Internetworking Technology Utilizing Satellite On-Board Switch," *Proceedings of the 20th International Communication Satellite System Conference*, May 2002; also AIAA Paper 2002-1964.

[5]Kadowaki, N., Yoshimura, M., Ogawa, Y., Komiya, N., and Araki, T., "Ka-Band Gigabit Communications Technology Satellite: Development Status and Connection Control Scheme," *Proceedings of the 5th Ka-Band Utilization Conference*, Taormina, Italy, Oct. 2002.

[6]Bever, M., Willoughby, S., Wiswell, E., Ho, K., and Linsky, S., "Broadband Payloads for the Emerging Ka-Band Market," *Proceedings of the 7th Ka-Band Utilization Conference*, Santa Margherita Ligure, Italy, Sept. 2001.

[7]Allman, M., "Enhancing TCP Over Satellite Channels using Standard Mechanisms," RFC2488, Jan. 1999.

[8]Yoshimura, N., Takahashi, T., Yoshikawa, M., Kadowaki, N., and Ikegami, T., "HDTV Transmission Experiment Using N-STAR," *Space Communications*, Vol. 14, No. 1, 1996, pp. 49–54.

[9]Kadowaki, N., Yoshimura, N., Nishinaga, N., Gilstrap, R., and Foster, M., "Trans-Pacific Demonstrations (TPD): Network Architecture, Engineering and Results," *Space Communications*, Vol. 17, No. 4, 2001, pp. 293–302.

Questions for Discussion

1) **What kind of antenna technology can be used for very high data rate transmission via satellite, and why?**

2) **List advantages and disadvantages of onboard switching, in particular considerations of hardware complexity (in space and on the ground), efficiency, and the various types of services that would benefit from onboard switching.**

3) **Describe the reasons why throughput performance of TCP is degraded when satellite link is used.**

4) **What kind of resolutions have been developed to avoid throughput degradation in TCP via satellite links?**

Chapter 4

Mobile Service Update

Hiromitsu Wakana*
Communications Research Laboratory, Tokyo, Japan

I. Introduction

MOBILE satellite communication systems are defined as radio-communication systems for communicating with terminals mounted on mobile vehicles such as ships, aircraft, and land vehicles using one or more satellites. The satellite could be either a geostationary satellite or low or medium Earth orbiting satellites (LEO/MEO). For the user links between mobile terminals and satellites, L-band, S-band, and Ku-band are now used, although higher frequencies, such as Ka-band, Q-band, V-band, and W-band, will soon come into use because these bands are less crowded and should make possible smaller user terminals. Using several experimental satellites such as the United States' ACTS,[1] Japan's COMETS,[2] Italy's ITALSAT,[3] mobile satellite communications experiments have been carried out in these higher frequency bands. For the feeder links between hub Earth stations and satellites, C-band, Ku-band, and Ka-band are now used.

The first-generation mobile satellite service (MSS) systems were characterized by global beam features of geostationary satellites and relatively large user terminals. The second-generation MSS systems play an important role wherever the terrestrial-based systems are not competitive, for example because of low traffic density in rural and remote areas, and to ease the terrestrial system during times of congestion. Third-generation MSS systems will use higher-frequency bands and make user terminals much smaller in size with higher capacity.

MSS systems are classified into geostationary earth orbit (GEO) and non-geostationary orbit (NGSO) satellite systems. The NGSO systems are further subdivided into LEO, MEO, and highly-inclined-Earth-orbiting (HEO) satellite systems. There are advantages and disadvantages of GEO, LEO, MEO, and HEO systems. The GEO systems have relatively simple configurations of both space segment and Earth stations, and extremely wide footprints on the ground, time-invariant look angles to the satellite, and a fixed propagation delay of

*Director, Yokosuka Radio Communications Research Center.

approximately 0.25 s for a single hop. However, they have power-limited links, troublesome propagation delays for voice services, and extremely low elevation angles in high-latitude countries.

In LEO systems, on the other hand, smaller, lower-power, and lower-cost user terminals are available, thanks to their lower propagation loss. Global services are also available, including even the Polar regions. However, they need a large number of satellites for continuous communications and have a shorter satellite visibility period, more frequent handoffs, and more complex onboard control systems. The altitude of LEOs is chosen to be in the range from 500 to 1500 km, which is below the two van Allen radiation belts at 1500 to 5000 km and 13,000 to 20,000 km. On the other hand, MEO satellites, which are positioned at altitudes of approximately 10,000 km between the inner and the outer van Allen belts, have intermediate characteristics between those of GEO and LEO satellites.

The HEO systems provide high elevation angles at high-latitude regions such as Europe and Canada, although they feature comparable propagation delays and path loss to those of the GEO systems. The Quasi-Zenith or Figure-Eight satellites,[4] which have the same altitude as that of GEO but have an inclination angle of approximately 45 deg, provide high-elevation angles as high as more than 70 deg for 24 h using three satellites even at medium-latitude countries like Japan. The most important feature is to ease requirements for mobile antennas like side-lobe radiation characteristics and their antenna tracking accuracy. However, they require large, tracking onboard antennas, and higher fuel consumption for satellite altitude controls.

Selection of the satellite orbit is a tradeoff between many factors—number of satellites, launch flexibility, minimum elevation angles within service areas, uplink and downlink frequencies, effects of van Allen radiation belts, handset power, propagation delay and loss, system reliability, and required service quality. The following sections describe operational and forthcoming mobile satellite communication systems including new aeronautical satellite communication systems and digital audio broadcasting services systems in the United States, Europe, and Japan.

II. Operational and Forthcoming MSS Systems

A. Inmarsat[5,6]

The first-generation global MSS was provided by the International Maritime Satellite Organization (Inmarsat). Inmarsat was established in 1979 as an international organization mandated with providing MSS for maritime users by leasing three Marisat satellites from COMSAT and two Marecs satellites from the European Space Agency (ESA). Table 4.1 shows relevant characteristics and data associated with existing GEO MSS systems and Table 4.2 shows follow-on Inmarsat procurement. In 1985, Inmarsat extended its service to include aeronautical services, and added land mobile services in 1989. In late 1994, Inmarsat changed its name to the International Mobile Satellite Organization. In 1999, Inmarsat was privatized into Inmarsat Holdings, Ltd., and its subsidiary company, Inmarsat, Ltd. To extend their business into very small aperture terminal (VSAT) and content solution providers, the name of the holding company was changed

Table 4.1 GEO satellites operating MSS

	Name							
	INMARSAT-3	N-STAR	AMSC 1	MSAT 1	Optus	Omnitracs/ Euteltracs	Garuda-1	Thuraya
Owner	Inmarsat	NTT DoCoMo	Mobile Satellite Ventures	Mobile Satellite Ventures	Optus	Qualcom	ACeS (Asian Cellular Satellite System)	Thuraya Satellite Telecommunications Co. Ltd
Country	UK	Japan	United States	Canada	Australia	United States	Asia	United Arab Emirates
No. of satellites	4	3	1	1	2	—	1	1
Latitude	54°W, 15.5°W, 64°E, 178°E	132°E, 136°E	101°W	106.5°W	160°E, 156°E	—	118°E	44°E
User frequency	L-band (1626.5–1646.5/ 1530–1545 MHz)	S-band	L-band (1646.5–1660.5/ 1545–1559 MHz)	L-band (1646.5–1660.5/ 1545–1559 MHz)	L-band (1646.5–1660.5/ 1545–1559 MHz)	Ku-band (14.0–14.5/11.7–12.7 GHz)	L-band (1626.5–1660.5/ 1525–1559 MHz)	L-band
Feeder link frequency	C-band (6.4/3.6 GHz)	C-band (6/4 GHz)	Ku-band (13/11 GHz)	Ku-band (13/11 GHz)	Ku-band (14.0115–14.0255/ 12.2635–12.2775 GHz)	Ku-band	C-band (6432–6725/ 3400–3700 MHz)	C-band
Launch vehicle	Proton, Atlas, Arian 4	Ariane from Kourou	Atlas Centaur	Ariane 4 from Kourou	Long March	—	Proton-Die	Sea launch
Launch date	1996	1996: -a & -b, 2002: -c	1995	1996	1992: B1& 1994: B3	—	2/13/00	10/20/00
Weight	2500 kg	2000 kg	2500 kg	2500 kg	1582 kg	—	N/A	3200 kg
Generation power	3000 W	5000 W	3000 W	3000 W	3500 W	—	N/A	13,000 W

Table 4.2 Inmarsat satellites

	Inmarsat-1			Inmarsat-2	Inmarsat-3
Spacecraft, launch date, and launcher	Marisat-F1: 2/19/76: Delta	Marecs-A: 12/20/81: Ariane	Intelsat V-MCS A: 9/28/82: Atlas-Centaur	F1: 10/30/90: Delta	F1: 4/3/96: Atlas Centaur IIA
	Marisat-F2: 6/10/76: Delta	Marecs-B2: 11/10/84: Ariane	Intelsat V-MCS B: 5/19/83: Atlas-Centaur	F2: 3/8/91: Delta	F2: 9/6/96: Proton D-1-E
	Marisat-F3: 10/14/76: Delta		Intelsat V-MCS C: 3/5/84: Atlas-Centaur	F3: 12/16/91: Ariane 4	F3: 12/18/96: Atlas Centaur IIA
				F4: 4/15/92: Ariane 4	F4: 6/3/97: Ariane 4 F5: 2/3/98: Ariane 4
Initial weight in orbit	330kg	560kg	1870kg	800kg	1133kg
Frequency					
L-band					
Satellite to Earth	1537.0–1541.0MHz	1537.5–1542.5MHz	1535–1542.5MHz	1525.5–1559MHz	1530–1648.1MHz
Earth to satellite	1683.5–1642.5MHz	1638.5–1644.0MHz	1636.5–1644MHz	1626.5–1660.5MHz	1626.5–1649.6MHz
C-band					
Satellite to Earth	4195–4199MHz	4194.5–4200MHz	4192.5–4200MHz	3599–3629MHz	3600–3623.1MHz
Earth to satellite	6420–6434MHz	6420–6435MHz	6417.5–6425MHz	6425–6454MHz	6425–6443.1MHz
Capacity	8 channels	40 channels	50 channels	250 channels	9*250 channels
max EIRP					
L-band				39dBW	40dBW/global 48dBW/spot
C-band				24dBW	27dBW

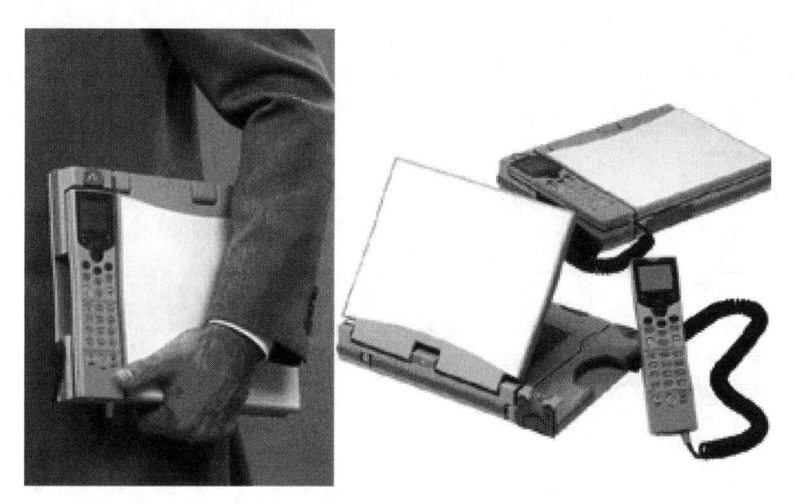

Fig. 4.1 Inmarsat Mini-M terminals (Courtesy of NEC).

from Inmarsat Holding, Ltd., to Inmarsat Ventures, Ltd. The services are provided using different types of Inmarsat standard terminals for maritime, land, and aeronautical mobile satellite communication services as shown in Fig. 4.1.

1. Inmarsat-A

The Inmarsat-A analog system provides direct-dial phone, data, telex, facsimile, and e-mail. Inmarsat-A parabolic antennas, which are approximately 1 m in diameter, are installed on various types of vessels, including oil tankers, liquid natural gas carriers, seismic survey vessels, fishing boats, and cargo vessels.

2. Inmarsat-B

Inmarsat-B, which is the digital replacement for Inmarsat-A and started its services in 1993, provides high-quality satellite phone, telex, medium- and high-speed data, and facsimile services with communication charges, lower than those for Inmarsat-A, ranging from $3 to $7 per minute. In May 2000, high-speed data (HSD) services for Internet access, data transmission, and distant medicine with images started.

3. Inmarsat-C

Inmarsat-C, initiated in 1991, provides a two-way, store-and-forward messaging, facsimile, and e-mail service (600 bps) via small terminals weighing only a few kilograms. The terminal can be mounted on a truck dashboard or at the helm of a small vessel, or carried around inside a briefcase. Inmarsat-C sup-ports the transmission of data such as location, speed, and heading, fuel stocks, and fuel consumption of vessels. It was approved for use under the Global Maritime Distress and Safety System (GMDSS). Since February 2000, e-mail service has also been available.

4. Inmarsat-D+

Inmarsat-D+ provides global two-way data communication and is suitable for supervisory control and data acquisition (SCADA) applications.

5. Inmarsat-E

Inmarsat-E provides global maritime distress alerting services transmitted from emergency position indicating radio beacons (EPIRB) and relayed through Inmarsat coast Earth stations.

6. Inmarsat-M

Briefcase-type Inmarsat-M communicators provide 6.4-kbps, two-way, digital voice and 2.4-kbps data and facsimile services. The combination of Inmarsat-M with the terrestrial wireless system is a cost-effective way of adding national and international reach to rural or remote areas where the telecommunication infrastructure is poor and there is a lack of satisfactory long-distance telephone links.

7. Inmarsat mini-M

The latest Inmarsat mini-M terminal is the smallest, lightest, and least expensive ever made. It is approximately 2 kg in weight, the size of a notebook computer, and has a cost of $2.50 per minute airtime. The use of such a small terminal has been made possible because of spot-beam coverage of the third-generation Inmarsat satellite. Maritime, land vehicle, and semi-fixed types are available. Voice, fax, and data services started in 1996. Since February 2001, using transportable terminals with high-gain antennas, a 64-kbps data transmission service named Inmarsat M4 has been in operation.

8. Inmarsat-Aero

The Inmarsat-Aero system provides interconnection service between jet aircraft and the public switched telephone network (PSTN). Using high-gain (12-dBi) steerable antennas, the system can provide 9.6-kbps digital voice and data services. A system using an omnidirectional antenna provides a low-bit-rate (600 bps) data service for aircraft fleet management. Several types of aeronautical services are available—Inmarsat-Aero-C, Aero-H, Aero-I, and Aero-L.

Inmarsat-Aero-C service is a messaging and data reporting service providing aircraft with store-and-forward satellite communications at 600 bps. This service is suitable for small aircraft in remote regions that do not need the full telephone and data capability provided by Aero-L and Aero-H. Inmarsat-Aero-H service provides aircraft with two-way digital voice, G3 facsimile at 4.8 kbps, and real-time data transmission at up to 10.5 kbps. Inmarsat-Aero-I provides telephony, G3 facsimile at 2.4 kbps, and real-time, packet-mode, data transmission at 4.8 kbps, via the spot beams of the Inmarsat-3 satellite. Inmarsat-Aero-L provides a two-way and real-time data transmission at 600 or 1200 bps using a low-gain antenna.

Recently, Inmarsat began a new service named Swift64, which gives aircraft passengers the ability to access Internet-based applications such as e-mail, video streaming, and file transfer with ISDN speeds of 64 kbps. Swift64 offers two types of data services—mobile ISDN and Internet protocol (IP)-based mobile packet data service. Swift64 avionics are compatible with the proven Inmarsat antennas and onboard infrastructure already installed on most of the world's long-haul airliners.

The 3 fourth-generation satellites with 100 times larger capacity than the third-generation satellites are being built by Astrium, Ltd. The Inmarsat I-4 network will deliver Internet and intranet content and solutions, video on-demand, video conferencing, fax, e-mail, voice, and LAN access at speeds up to 432 kbps.

B. Japan's N-STAR[7]

After mobile satellite communications experiments using Engineering Test Satellite Five (ETS-V) and ETS-VI, two N-Star satellites were launched into geostationary orbit slots of 132° E and 136° E by Ariane rockets in Kourou, French Guiana, on 29 August 1995, and 5 February 1996, respectively. Using two S-band transponders in each satellite, commercial mobile satellite communication services to portable terminals and land vehicles in domestic areas and to vessels up to 370 km offshore from Japan started on 29 March 1996 by NTT DoCoMo. DoCoMo is a cellular service provider that targets the extension of cellular services to areas not covered by cellular systems and the extension of maritime two-way communications both onshore and offshore. On the N-STAR satellites, the antenna gain has been increased using a multibeam antenna covering the Japan islands and closeby offshore regions with four spot beams. This allows a reduction in the size of user terminals. Moreover, N-STAR c was successfully launched by Ariane 5 on 6 July 2002 from Kourou, French Guiana, into a geostationary orbit of 136°E.

By using N-STAR a and N-STAR b an MSS named "Wide Star" is being provided, including satellite portable and car phone services, satellite maritime phone, satellite packet communication, and satellite airplane phone services. N-STAR c is expected to give more reliability to Wide Star services, including next-generation mobile satellite services such as disaster communications, multicast, and broadcast services.

The satellite portable phone terminal has two modes—satellite phone and terrestrial cellular phone. When a user is inside the service area of cellular systems, users use cellular mode, but when a user is outside the cellular network in rural and remote areas, it selects the satellite mode. In addition to telephony, service applications being provided include mass communications for reporters, allowing them to send photos and messages from remote areas, as well as similar services to manufacturing companies, mining and exploration companies, construction companies, and life-saving emergency care and data acquisition and transmission in the event of natural disasters. Figure 4.2 shows several terminals for N-Star services.

Fig. 4.2 N-Star terminals.

C. United States' AMSC[8] and Canada's MSAT[9,10]

In 1989, the Federal Communications Commission (FCC) licensed the American Mobile Satellite Consortium (AMSC), which was formed from eight of twelve applicants for the first-generation domestic MSS services in the United States. AMSC later changed its name to American Mobile Satellite Corporation, whose major shareholders include Hughes Communication, Inc., AT&T Wireless Services, Inc., and Singapore Telecom.

On the other hand, the Canadian government commenced the MSAT program at the Communication Research Centre in Ottawa. In 1988, the MSAT project moved from the government to the private sector under Telecom Mobile, Inc. After going bankrupt in 1993, TMI was reconstructed as TMI Communications & Company, Ltd. (TMIC), with investment by Bell Canada Enterprise.

AMSC and TMIC signed contracts with Hughes and Spar Aerospace, Ltd., in 1990 to build their respective satellites, operate identical spacecraft, and provide complementary mobile services with backup and restoration for each other. The first AMSC satellite, AMSC-1, was launched by an Atlas IIA rocket from Cape Canaveral on 7 April 1995, and MSS services began in 1996. The MSAT-1 was launched by an Ariane 4 from Kourou, French Guiana, on 20 April 1996. AMSC-1 is located at 101° W, and MSAT-1 is at 106.5°W. Each satellite has the capacity to support up to 2000 simultaneous radio channels.

For user links, 28 MHz of L-band spectrum (1545 to 1559 MHz and 1646.5 to 1660.5 MHz) was allocated, and for the feeder link, 200 MHz in the Ku-band was allocated. Equivalent isotropic radiated power (EIRP) in L-band is 57.3 dBW and the EIRP in Ku-band is 36 dBW. Each satellite uses four spot beams at L-band to cover North America and coastal areas up to 300 km offshore. Another beam covers Hawaii and Alaska. Each satellite is approximately 18.9 m from side to side, when its two L-band antennas are deployed, and 21 m high from the tip of one solar wing to the tip of the other. The 6.8-m × 5.25-m elliptically shaped L-band antennas are made of graphite and weigh only 20 kg each. Instead of being folded against the spacecraft body for launch as are conventional antennas, these antennas are rolled together into a 4.9-m-high cone shape atop the satellite. The cone is approximately 1.5 m in diameter at the top and 3 m in diameter at the bottom.

AMSC acquired the ARDIS company in 1998, to expand their product line with ARDIS's terrestrial network, and changed its name to Motient in 2000 to solidify the consolidation of the two companies. In November 2001, the satellite

Fig. 4.3 AMSC's land mobile satellite phone (Courtesy of Mitsubishi Electric Corporation).

division of Motient Corporation and Telesat Mobile, Inc., Communications joined to form a new company, Mobile Satellite Ventures (MSV). Figure 4.3 shows an example of a land mobile satellite telephone for the MSV systems. Full duplex digital voice, G3 facsimile and data services at 4800 bps are available. The medium-gain antenna shown in Fig. 4.3 is 1.4 kg in weight, 170 mm in diameter and 165 mm in height with a terminal performance of a G/T of −16 dB/K and an EIRP of 12.0 to 16.6 dBW.

D. Australia's MobileSat[11,12]

Mobile satellite communication service, named "MobileSat" is provided by Optus Communications Pty, Ltd., to Australia and its neighboring waters up to 200 km. In August 1992, the OPTUS-B1 satellite was launched by a Long March rocket in China. After an unsuccessful launch of the B2 in December 1992, the B3 satellite was successfully launched in August 1994. In 2009, the B3 satellite is expected to enter into inclined orbit, ensuring the provision of L-band MSS services until 2014. Operating through the OPTUS-B1 and OPTUS-B3 satellites, the MobileSat provides secure, reliable, direct-dial telephone, fax, data, and messaging services. In 2001, SingTel became the parent company of Optus to become a strong and strategic telecommunications player within the Asia-Pacific region.

The mobile users use right-handed circularly polarized L-band signals, in the 1646.5 to 1660.5 MHz band for the uplink and 1545.0 to 1559.0 MHz for the downlink. With a nominal transmission power of 150 W for each of their L-band transmitters, the two OPTUS-B satellites, located at 156° E and 160° E, provide an EIRP of 46 dBW over Australia. The uplink L-band signals are converted to Ku-band at 12.2635 to 12.2775 GHz and are relayed to one of two network management stations (NMS) in Sydney and Perth. For the Ku-band uplink, 14.0115 to 14.0255 GHz is used. Access to the public switched telephone network (PSTN) is via a gateway station, which is controlled by the NMS. The NMS is also responsible for overall management and coordination of network operations.

The Optus MobileSat units can be mounted to land vehicles, maritime vessels, aircraft, and portable cases and provide digital voice, fax, and data transmission. Circuit-switched, full-duplex, digital-voice communication between a fixed phone, a cellular telephone, and MobileSat telephone are provided by the MobileSat service. For examples of service charges, fixed line to MobileSat calls, MobileSat to MobileSat calls, and MobileSat phone to cellular phone cost $1.80, $2.96, and $2.20 per minute, respectively. Two types of MobileSat handsets are available, manufactured by Westinghouse and NEC.

E. OmniTRACS and EutelTRACS[13]

In 1988, Qualcomm, Inc., introduced land mobile communication services using Ku-band of 14/12 GHz as a user link in North America. The target market was the long-distance road haulage industry. Now, more than 1400 United States fleets use the OmniTRACS to communicate with drivers, monitor vehicle location, and provide customer services. More than 425,000 units were shipped worldwide. The OmniTRACS mobile communications system provides real-time ubiquitous satellite tracking, two-way data communication, and complete dispatch and back-office integration capabilities. OmniTRACS can improve the security of shipping goods, including sensitive cargo, by truck across the United States. Broad satellite coverage, coupled with GPS, enables trucking companies to determine the location of their vehicles and communicate with the drivers at all times.

The OmniTRACS steerable antenna mobile terminal is 10 in. wide, and has an azimuth beamwidth of 6° and an elevation beamwidth of 40° and with an antenna gain of 19 dBi. The data rate is 600 bps to and 165 bps from mobile terminals. The transmitting power from the mobile terminal is 1 W. Direct sequence spread spectrum is used to reduce the signal power density by spreading the signal spectrum over 2 MHz. In the forward link from the hub to mobile terminals, time division multiplexing is used.

According to statistics of the United States' MSS market, total subscribers were almost 350,000 as of 1997, split between several MSS providers (60% for OmniTRACS, 28% for Inmarsat, 9% for AMSC, and 3% for others). The majority of subscribers are using data-only messaging services (data service is 75% and voice service is 25%).

In 1991, this service was introduced to the European market as EutelTRACS by using Eutelsat Ku-band satellites. As the first European commercially operated mobile satellite service, EutelTRACS provides two-way mobile satellite communication and vehicle positioning services dedicated to enhance fleet management productivity. In Europe, there are several service providers, including Alcatel SEL in Austria and Germany; Alcatel Mobicom in Belgium, Luxembourg, and the Netherlands; D&COMM in Czech Republic; Knud Hansen Kommunika-tion in Denmark and Sweden; Alcatel Mobicom France in France and Portugal; Antenna Hungaria Corporation in Hungary; Telespazio S.p.A. in Italy; and Ronda Grupo Consultor in Spain.

F. ACeS[14]

Bermuda-based ASIA Cellular Satellite (ACeS) International owns the ACeS system. Their primary shareholders are Indonesia's Pasifik Satelit Nusantara,

Lockheed Martin Global Telecommunications, the Philippines Long Distance Telephone Company, and Thailand's Jasmine International Overseas Company, Ltd. AceS can provide voice, facsimile, and pager services with handheld and fixed terminals to nearly 3.5 billion people, approximately 60% of the world's population, over the whole of Asia including Southeast Asia, India, Pakistan, Papua New Guinea, China, Korea, and Japan. The ACeS satellite, named "Garuda-1," was developed by Lockheed Martin Commercial Space Systems and launched on 13 February 2000 by a Proton-D1e rocket from the Baikonur Consmodrome in Kazakhstan. Mission life is designed to be 12 years.

The ACeS is a geostationary satellite located in 118°E. The user link between mobile users and the satellite uses L-band between 1626.5 to 1660.5 MHz in the uplink and 1525 to 1559 MHz in the downlink. With 140 spot beams with 20-fold frequency reuse and a beam isolation of 14 dB, AceS can provide 73 dBW of EIRP and 15.3 dB/K of G/T. The feeder link uses C-band at 6432 to 6725 GHz and 3400 to 3700 MHz. One of the key elements is the power amplifier system, which uses a multiport amplifier where several high-efficiency, solid-state power amplifiers are combined coherently to deliver RF power to individual beams as a function of user demand. Thanks to the large separate L-band antennas, and an onboard switching capability, L-band-to-L-band links between mobile users are also available.

Table 4.3 shows the capacity of available communication links between mobile users and gateways. In the C-to-L link, (C-band uplink from a gateway Earth station and L-band downlink to mobile users), a maximum of 896 frequency division multiplexed (FDM) channels are available, and the basic voice service has 32 duplex calls multiplexed per channel. Therefore, a total of 28,672 calls can be handled simultaneously.

The ACeS L-band, multibeam antenna consists of two 12-m diam, 7.8-m-focal-length, offset reflectors located on the East and West sides of the spacecraft. A planar feed array is composed of 88 crossed dipole elements in hexagonal cups. The 140 communication beams, as well as the additional 8 beams used to adjust the antenna pointing, are formed by controlling the amplitude and phase of each individual element using two complex beam forming networks.

An example of an ACeS terminal is the Ericsson R190. It has dual-mode capability of satellite and GSM 900 mobile phone, which provides voice and data 2400 bps (satellite) and 9600 bps (GSM). Whenever a GSM network is not

Table 4.3 ACeS channel capacity

Link	Channel capacity
C to L links	896 FDM channels 32 duplex calls multiplexed per channel
L to C links	896 FDM channels 32 duplex calls multiplexed per channel
C to C links	78 FDM channels
L to L links	96 FDM channels 32 duplex calls multiplexed per channel

available, the mobile phone immediately switches over to satellite operation. This terminal weighs 210 g, and is 130 × 50 × 32 mm in size. Standby time is 43 h and talk time is 3 h 40 min in both modes.

G. Thuraya

Thuraya Satellite Communications Company is a private joint stock company established in the United Arab Emirates (UAE). The Thuraya satellite was built by Boeing Satellite Systems (BSS) under a contract signed on 11 September 1997. It was launched on 20 October 2000 into a geostationary orbit at 44°E. The coverage area encompasses 99 countries over Europe, North and Central Africa, the Middle East, Central Asia, and the Indian Subcontinent, encompassing nearly 40% of the world's population. Thuraya provides GSM-compatible mobile phone services using a single satellite antenna with a 12.25 m × 16 m mesh reflector.

Fig. 4.4 Thuraya terminals, Ascom and Hughes (Courtesy of Thuraya Satellite Telecommunications Company).

Using onboard digital signal processing, the antenna creates more than 200 spot beams that can be redirected onboard to adapt business demands in real time. Up to 13,750 simultaneous voice circuits are available.

Figure 4.4 shows a Thuraya terminal with a dual-mode satellite/GSM mobile phone that has been developed by partnership with Ascom and Hughes Network Systems and which works both in GSM networks and in satellite networks. The terminal, which weighs 220 g with quadrifilar helix L-band antenna, provides voice, fax, data, and short message services (SMS) at 9600 bps. Standby time is 34.1 h in satellite mode and 33.3 h in GSM mode. Talk time is 2.4 h in satellite and 4 h in GSM.

H. Other Proposed MSS

Agrani Satellite Services has put its mobile-telephone satellite on hold because of poor market conditions. The company has shifted its focus toward a conventional telecommunications satellite above India with C-band and Ku-band. Asia Pacific Mobile Telecommunications (APMT) was planned to provide personal satellite communication services using handheld terminals and land mobile satellite communications services. Two large GEO satellites were expected to be launched in 1998, but in July 1998 the export license for Hughes, which developed the satellites, to export the APMT satellite to China, was suspended by the U.S. government.

I. Big LEO Satellite Systems

In 1990 and 1991, six main contenders for big LEO systems filed with the FCC to provide global satellite service: Iridium, Odyssey, Globalstar, Ellipso, Aries, and AMSC. On 31 January 1995, three systems—Iridium, Globalstar, and Odyssey—were awarded licenses by the FCC to operate in the United States. Industry experts believed that only one or two of these systems could survive. However, the handheld market for the big LEO systems is not the large market that was expected, and the actual demand was only approximately 10% of the predictions. On the other hand, the terrestrial cellular phone market was much larger than was expected. This may be in part because satellite services are not available inside of buildings (therefore users have to go outside to use satellite systems), yet cellular phone services are available inside buildings and even in many underground locations and subways.

On 17 July 2001, the FCC granted spectrum licenses to eight new MSS systems in 70 MHz of the 2-GHz band (1990–2025 MHz and 2165–2200 MHz) to provide domestic service in the United States. The authorizations were issued to The Boeing Company; Celsat America, Inc.; Constellation Communications Holdings, Inc.; Globalstar LP; ICO Services, Ltd.; Iridium LLC; Mobile Communications Holding, Inc.; and the TMI Communications and Company LP. All the licensees must have physical construction of all satellites underway by January 2004 with the entire system certified by July 2007. These eight companies have proposed practically every possible constellation of satellites, ranging from LEO, MEO, to GEO and mixed.

Table 4.4 Characteristics of big LEO

Name	Owner	No. of satellites	Altitude, km	Period, min	Inclination, deg.	No. orbit planes	Weight, kg	Generation power, W	User frequency, MHz	Multiple access
Iridium	Motorola, Inc./Iridium, Inc.	66 + 6 spares	785	100.13	86.4	6	689	1200	1616–1626.5	TDD/TDMA/FDMA
Globalstar	Loral/Qualcomm	48 + 8 spares	1414	113	52	8	426	1000	1616–1626.5 2483.5–2500	CDMA
ICO	ICO Global Communications, Ltd.	10 + 2 spares	10,355	360	45	2	1600	5100	1980–2010 2170–2200	TDMA

Table 4.5 Five 2nd-round 2-GHz LEO

System	Owner	Constellation
Boeing Co.	Boeing Co.	16 MEOs
GS-2	Globalstar LP	64 LEOs and 4 GEO
Macrocell	Iridium LLC	96 LEOs
Ellipso 2G	Mobile Communication Holdings, Inc.	26 MEOs
Constellation 2	Constellation Communications, Inc.	12 ECCO + 46 global

Table 4.4 shows the initial plans of big LEO systems (Iridium, Globalstar, and ICO) and Table 4.5 shows the 5 second-round, 2-GHz LEO systems licensed by the FCC. In this section, several big LEO systems will be described.

1. *Iridium*[15,16]

The Iridium network consists of a constellation of 66 satellites, 785 km in altitude, with 6 polar orbital planes inclined 86.4 deg, each containing 11 satellites with an orbital period of approximately 100 min. Each satellite uses 3 L-band antennas to cover the ground with 48 beams, and the diameter of each spot beam is approximately 600 km. The 66 satellites provide 3168 cells, of which only 2150 need to be active to cover the whole surface of the Earth. Uplink and downlink frequencies are identical in the range from 1610 MHz to 1626.5 MHz. Using 50-kbps time division multiple access (TDMA) bursts in uplink and downlink, 4800-bps voice or 2400-bps data, full-duplex communications services are available. Ka-band intersatellite links with four crosslinks on each satellite provide high-speed communications between neighboring satellites.

In November 1998, Iridium began providing commercial services, but less than 1 year later the company filed for bankruptcy in August 1999. Iridium services were terminated on 17 March 2000, after the company failed to find a rescuer from bankruptcy, although Motorola maintained the network for a limited period of time for users in remote areas. The U.S. Bankruptcy Court for the Southern District of New York gave Motorola permission to deorbit the satellites and to have them reenter into the atmosphere to avoid becoming debris that could be dangerous to other satellites.

The Department of Defense (DoD), however, subsequently signed a $72 million contract with Iridium Satellite LLC for 24 months of MSS services. This contract would provide unlimited airtime for 20,000 government users of satellite phones including military forces worldwide. The contract includes options that could extend the service period to December 2007. Iridium Satellite LLC bought all of the existing assets of Iridium LLC, including its constellation of satellites and its satellite control network, and had Boeing operate the system.

Iridium's original phones cost as much as $2000 to $3000 with airtime fees of $3 to $7 per minute. Instead, Iridium Satellite plans to target industrial business markets, such as aviation and oil and gas exploration concerns, in addition to government customers.

2. *Globalstar*[17]

The Globalstar system has a constellation of 48 satellites in 8 planes with 6 satellites per plane inclined at 52 deg, which was chosen to provide 100% coverage from 70° N to 70° S. The orbital altitude is 1414 km and the orbital period is 114 min. The user link uses the L-band at 1610 to 1626.5 MHz for uplink and the S-band at 2483.5 to 2500 MHz for downlink. The feeder link uses the C-band at 5 GHz for uplink and 7 GHz for downlink. The L- and S-band satellite antennas are active phased array antennas, which divide the user coverage into 16 beams that collectively fill the 5760-km-diam circle on the Earth visible to an individual satellite. The channel capacity of one satellite is 2148 links. The satellite transponders are not regenerative but simple bent-pipe types without intersatellite capability. The satellites use code division multiple access (CDMA) with an efficient power control technique, multiple-beam, active-phased array antennas for frequency reuse, variable rate voice encoding, multisatellite path diversity, and soft handoff for both beams and satellites.

The first and second lots of four Globalstar satellites were launched on 14 February 1998 and on 24 April 1998, respectively, by Boeing Delta 2 rockets from Cape Canaveral. On 9 September 1998, however, the third launch, of 12 satellites from Baikonur Cosmodrome in Kazakhstan, failed because of a flight control computer problem. Launches number 4 through 13 took place during 1999 by Soyuz-Ikar and Boeing Delta rockets, placing another 40 satellites into orbit. The fourteenth launch of four remaining satellites as in-orbit spares, was carried out on 8 February 2000 by a Delta 2 rocket, bringing the total number of Globalstar satellites in orbit to 52.

Globalstar commenced commercial services in 2000, and as of 15 February 2001, Globalstar was providing MSS services via 25 gateways that cover 109 counties. As of 31 December 2000, Globalstar had 31,000 subscribers, and by June 2002, the number had increased to 66,000 subscribers.

On 15 February 2002, however, Globalstar filed a voluntary petition under Chapter 11 of the U.S. Bankruptcy Code in the U.S. Bankruptcy Court in Delaware. The company is drastically cutting costs to reduce a gap between revenues and expenses, and has put in place a new business model designed to secure larger customers in industries that need remote communications. Globalstar was granted the use of 2 GHz spectrum as one of the new applicants for the second-generation MSS systems.

3. *ICO*

In January 1995, ICO Global Communications, Ltd., was established as a private spinoff company from Inmarsat to provide personal and mobile satellite communication services. The ICO space segment was to consist of ten operational satellites and two in-orbit spares at an altitude of 10,354 km, which is an intermediate circular orbit (ICO). Two orbital planes, inclined 45° to the equator, are used to provide continuous overlapping coverage of the ground. The footprint diameter on the ground with minimum elevation angle of 20° is 10,850 km and the maximum footprint pass time is approximately 97 min. Using TDMA, each satellite can handle 4500 simultaneous telephone calls.

After Iridium filed for bankruptcy in August 1999, ICO also filed for bankruptcy on 27 August 1999. ICO emerged from Chapter 11 after Craig McCaw led a group of international investors to provide $1.2 billion to acquire the ICO business in May 2000. In 2002, ICO agreed to join with two other MSS operators—Constellation Communications Holdings Inc. and Mobile Communications Holdings Inc. (MCHI)—both licensees in the 2-GHz band, and was endeavoring to initiate next-generation mobile satellite operations.

The new ICO system design was updated to meet higher data rate and Internet services in addition to the original MSS satellite phone. The ICO satellites are modified versions of the Boeing 601, and are 25% taller because of the size of their transmit and receive antennas. With a power of 5100 W and an EIRP of 58 dBW, 4500 simultaneous telephone calls are available.

The second launch of the ICO F-2 satellite was successfully done by an Atlas 2AS rocket on 19 June 2001, although the first launch from Sea Launch failed on 12 March 2000. Tests for evaluating the orbiting satellite and ground infrastructure will be carried out. Commercial services will be started in 2003. Potential markets are maritime, aviation, oil and gas, transportation and construction industries, government agencies, and individual consumers.

J. Little LEO Satellite Systems

The FCC has licensed three little LEO systems—Orbcomm, Starsys, and Volunteers in Technical Assistance (VITA) in the frequency bands of 137 to 138 MHz and 148 to 149.9 MHz. In November 1994, five additional companies filed applications for licenses for new little LEO systems with the FCC (CTA, LeoOne, Final Analysis, E-Sat, and GE-Americom). These little LEO systems use the very/ultra-high frequency VHF/UHF band, because it allows user terminals to be made at lower cost because these frequencies are already very well used. The drawback, however, is the same; these frequencies are heavily used worldwide for public and private services. Finally, almost all the little LEO systems except Orbcomm are not operating actively, although VITA is operating on a voluntary nonprofit basis. Table 4.6 shows the characteristics of little LEO systems (Orbcomm and VITASAT).

The Orbcomm system was developed by Orbital Communications Corporation, a subsidiary of Orbital Sciences Corporation, in cooperation with Teleglobe Canada, Inc., to provide low-cost two-way data and message communications. In the first phase of operation, there was a constellation of 26 satellites, which included 24 satellites in 3 orbital planes of 8 satellites, each with an inclination of 45 deg at 775 km altitude. In the second phase, a fourth plane of satellites at a 45-deg inclination and a second polar orbit with 2 satellites was added, totaling 36 satellites in orbit. Uplink from users is 2400 bps at 148 MHz, and downlink to users is 4800 bps at 137 MHz. Orbcomm provides short message services of 6250 bytes typical and location services to approximately 40,000 mobile terminals. Figure 4.5 shows an Orbcomm satellite.

On 15 September 2000, the company filed for bankruptcy protection. By that time Orbital Sciences, Teleglobe, and bond holders had invested $810 million into

Table 4.6 Characteristics of little LEO

Name	Owner	No. of satellites	Altitude, km	Inclination, deg	No. orbit planes	Weight, kg	User frequency, MHz	Feeder link frequency, MHz	Multiple access
Orbcomm	Orbital Comms Corp.	32/4	775/739	45/70	4/2	38.5	148.905–159.9 (up) 137–138 (down)	149.61/137.56	FDMA (user), TDMA (feeder link)
VITASAT	Volunteers Technical Assistance	2	970	88	1	45	148.055–150.395 (up) 137.09–137.9 (down)	400.28–400.91	FDMA

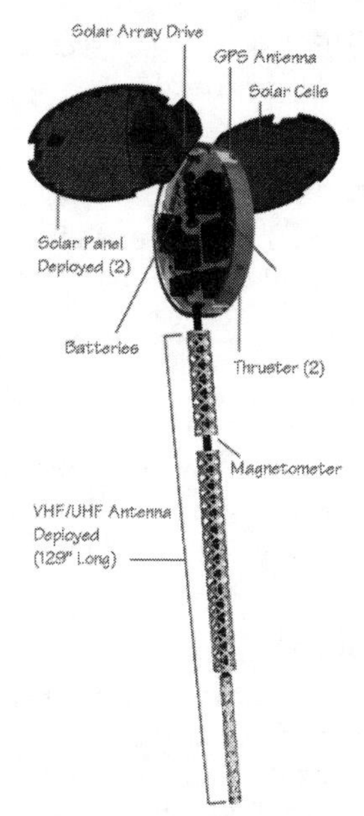

Fig. 4.5 Orbcomm satellite (Courtesy of ORBCOMM).

the company. On 9 March 2001, Advance Communications Technologies was selected as a new owner following a managed bankruptcy auction for Orbcomm.

III. Aeronautical Satellite Communications[18]

Since the Inmarsat started aeronautical satellite communication services in 1991 by using L-band, voice communication for passengers and pilots in aircraft is drastically improved. Services for delivering e-mail and Internet access as well as news, financial data, and sports and weather information to airborne passengers are now available. Business travelers, who make up half of all fliers, number approximately 40 million annually worldwide, and they have to stay in the cabin for several hours without any interaction to the business world.

Several companies (such as Tenzing Communications, Inc., and Airshow, Inc.) are providing Internet access and e-mail services for passengers at low-data rates such as 2.4 to 9.6 kbps. They are planning to enhance their services using the new higher data rate (64 kbps) L-band channels of Inmarsat. On the other hand, AirTV is proposing another solution of S-band using new S-band geostationary satellites to provide global and higher data rate services including TV programs. Boeing will start earlier services of broadband aeronautical communications using existing Ku-band satellites, although the Ku-band is not allocated to Aeronautical Mobile Satellite Services (AMSS) by the Radio Regulation. In the next 2003 meeting of World Radio Communication Conference (WRC), the new allocation of Ku-band to AMSS will be discussed. In this section, these aeronautical satellite communications systems will be described.

A. Tenzing Communications[18]

Tenzing Communications Inc. has been providing narrowband (low data rate) Internet access and in-flight e-mail services since 2001. Passengers connect their laptop to an onboard server via the seat-back phone port, and can send e-mail at the average connection rate of 56 kbps to the server. The e-mail does not transfer to the ground servers in real time, but are sent every 15 min to the ground at 2.4–9.6 kbps after being compressed. Despite the 15-min delay, passengers will not be aware of the actual transmission of their e-mail. They are planning to provide a new bidirectional service using a higher-speed channel of Swift64 of Inmarsat at 64 kbps. Using two channels, 124 kbps services are also available.

B. Airshow/Rockwell Collins[18]

Airshow Inc. who is a provider of cabin management and passenger information systems to both business aviation and commercial airlines, has provided narrowband real-time data communications services. In April 2002, Airshow announced a broadband terminal working at 144 kbps of bidirectional data transmission over Globalstar's LEO system. After Rockwell Collins, Inc. acquired Airshow in June 2002, they introduced Collins Airshow 21, which is a natural extension of their previous avionics systems. To provide a high-speed and bidirectional voice/data transmission of 64 kbps, 60-W high power amplifier (HPA), traditional high-gain antenna, Inmarsat's Aero-H/H+ and Swift64 capabilities are used. Intracabin data communications are achieved with a wireless LAN of the IEEE 802.11b standard, allowing passengers to connect with the onboard server from anywhere in the cabin. Rockwell Collins signed a memorandum of agreement with Airbus to provide the Aircraft Integrated Network System (AINS) for the A340-600, and later they expanded to include the Airbus in-Flight Information Services (AFIS) for the A340 and A320 family. In July 2002, Airbus, Rockwell Collins, and Tenzing Communications, Inc., announced cooperation to offer aeronautical communication services, and at the third phase of the development will add broadband aeronautical satellite communication services, including television, news, live sports, and weather and business information.

C. AirTV

AirTV plans to build a new global satellite network by using four S-band geostationary satellites located at 86° W, 12° W, 64° E, and 180° E. Each satellite will deliver 60 digital TV channels as well as e-mail, Internet access, and airline communications, with 80 Mbps from four transponders. The AirTV system is being developed by a team of Alcatel Space (satellites and ground system), Arianespace (launch services), CMC Electronics (aircraft antenna), and SITA (return link for data services) under strategic alliances with Tenzing Communications, Inc., and Rockwell Collins. The service will be started in late 2004 and will be fully global in

2005. The S-band of 2.5 GHz was selected because of global allocation of ITU, suitable orbital positions found, and sufficient bandwidth for AirTV service.

D. Connexion by Boeing[18]

In April 2001, the FCC granted Boeing permission to operate 800 receive-only aeronautical Earth stations at 11.7–12.2 GHz, under the condition that the stations not cause harmful interference to other allocated services in this frequency band. Only one-way service to passengers and crew aboard aircraft is allowed. The Telstar-6 and GE-4 satellites were used within the United States for this test trial. In December 2001, Boeing was granted a license by the FCC to both receive and transmit broadband data in flight. The Connexion by Boeing system is aimed at nonsafety communications such as passenger entertainment and passenger/crew communications to ground using existing Ku-band satellites.

The system consists of a space segment, which leases the capacity of the existing Ku-band fixed satellite service (FSS) satellites, AMSS terminals, ground Earth stations (GES) for FSS, and a network operations center (NOC) for avoiding interference to other systems. Multiple forward links from GES to AMSS have a transmission rate of 4.86 Mbps, and return links from AMSS have 16 kbps to 1024 kbps sharing one transponder with CDMA.

After an AMSS terminal receives navigation data from the airborne navigation system, tracks the antenna to an expecting satellite, and can receive the forward-link signal, the AMSS terminal must begin to transmit an initial acquisition signal at 16 kbps. Depending on user traffic, interference level to the other systems, and C/N_0 (carrier power-to-noise power density ratio) at the AMSS receiver, EIRP and data rate are controlled. The NOC controls the total EIRP from all operating AMSS to keep the total signal power less than the regulation level of ITU-R Recommendation 728-1.

The AMSS uses separate transmitting and receiving antennas using active phased array techniques, attached on the top of the fuselage separated 1.25 m from each other. A specification example of the AMSS transmit antenna is 38 cm in diameter, 873 elements, an EIRP of 51.2 dBW in boresight, a beamwidth of 3.2 deg by 3.5 deg, a maximum scan angle of 63 deg, and linear polarization. A receive antenna has a diameter of 43 cm by 61 cm, 1515 elements, a G/T of 12.5 dB/K in boresight, a beamwidth of 2.1 deg by 3.1 deg, a maximum scan angle of 63 deg, and a circular polarization. As the antenna beam is scanned from the boresight, the antenna gain decreases by a cosine function, and the beamwidth increases by an inverse cosine function.

The closed-loop antenna tracking method is used; the transmit antenna tracking is slaved with the receive antenna tracking. The 50-Hz step tracking achieved a pointing error of 0.08 deg in the test flight. The total pointing error including all the error factors will be less than 0.15 deg in calculation. In both forward and return link, direct-sequence spread spectrum technique of an offset QPSK is used.

E. Regulatory Status[19]

The secondary allocation to MSS was added by the WRC-95 to introduce land mobile satellite services in the Ku-band by OmniTRACS and EutelTRACS, but

Table 4.7 Regulatory issues in the 14.0–14.5 GHz band

	Frequency, GHz					
	14	14.25	14.3	14.4	14.47	14.5
Primary services						
Fixed satellite service			Earth-to-space			
Fixed service		S5.505	R1&R3		All regions	
		S5.508				
Mobile service		S5.509	R1&R3		All regions	
Radio navigation service		all region				
Secondary services						
Mobile satellite service			Earth-to-space except AMSS			
Space research and radio astronomy services		Space research		Space research	Radio astronomy	
Radio navigation service			all region			

aeronautical mobile satellite was excluded, because any aeronautical application plans were not considered at that time.

In the WRC-03, the technical and operational studies evaluate the feasibility of sharing the band of 14–14.5 GHz between the AMSS and other services such as fixed-satellite, radio navigation, fixed and mobile service, radio-astronomy service with the former on a secondary basis. The major responsibility of this subject is assigned to WP8D (all mobile satellite services). Working groups of WP4A (efficient orbit/spectrum use of fixed satellite service), WP9D (sharing with fixed service and other services), WP7B (space radio system), WP7D (radio astronomy), and WP7E (interservice sharing and compatibility of science services) are requested to support this job. Table 4.7 shows the frequency allocation in 14.0 to 14.5 GHz in three regions—R1 for Europe, R2 for the Americas, and R3 for Asia and Pacific with footnotes of Article S5.

In terms of sharing with the FSS, because the currently planned AMSS systems will use the exiting Ku-band satellites, interference from the AMSS terminals on a secondary basis to the adjacent FSS satellites operating at the same frequency band of 14–14.5 GHz will cause the most problems.

The WP4A has concluded that the current regulatory restriction on the use of AMSS in 14–14.5 GHz is simply removed, and emission from either an AMSS terminal or a fixed satellite terminal is not distinguished in term of interferences toward adjacent satellites. Mobility of AMSS terminals is not of concern, but interference level for AMSS terminals must meet the same restrictions as VSAT's conditions, S728-1.

IV. Digital Audio Broadcasting Services

Digital broadcasting services of CD-quality music programs from satellites to listeners in mobile vehicles and at home have been attracting attention. This service, generically referred to as digital sound broadcasting (DSB), is also known by several other names: digital audio broadcasting (DAB) internationally, digital audio radio services (DARS) in the United States, and digital radio services (DRS) in Canada. Several hundred satellite radio channels operating in the Ku-band at 10–11 GHz or 12 GHz are available all over the world, but these services are assumed to be provided to fixed receivers that have a line-of-sight path to the satellite. However, digital audio broadcasting systems using L-band or S-band are designed for mobile and portable reception.

The ITU's World Administrative Radio Conference for Dealing with Frequency Allocations in Certain Parts of the Spectrum (Malaga-Torremolinos, in 1992, or WARC-92) allocated the following frequency bands to the satellite sound broadcasting service and complementary terrestrial broadcasting services for the provision of digital audio broadcasting:

1) 1452 to 1492 MHz worldwide, except for the following specific countries;
2) 2310 to 2360 MHz for the United States and India; and
3) 2535 to 2655 MHz for Japan, the Republic of Korea, China, the Russian Federation, the Ukraine, Belarus, Singapore, Thailand, Pakistan, Bangladesh, and Sri Lanka.

The worldwide allocation is the L-band (1.5 GHz), but the United States and a group of countries mainly in Asia selected S-band.

In the following sections we will introduce two U.S. satellite radio systems: the XM Satellite Radio system and the Sirius Satellite Radio system. In 1992, the FCC allocated S-band for nationwide DARS, but only four companies applied for a license to broadcast over that band. The FCC gave licenses to two of these companies in 1997—CD Radio (now Sirius Satellite Radio) and American Mobile Radio (now XM Satellite Radio).

A. XM Satellite Radio[20,21]

The XM Satellite Radio (XMSR) system, which is a satellite-based digital audio broadcasting system operating in the United States, has a space segment consisting of two satellites: "Rock" located at 115° W and "Roll" located at 85° W.

Table 4.8 shows the frequency and the polarization of downlink signals to subscribers from the satellites. Each of two ensembles A and B contains half of the

Table 4.8 Downlink frequency and polarization of the XM Satellite Radio System

Ensemble A			Ensemble B		
Roll A1	Rock A2	Terrestrial Repeater At	Terrestrial Repeater Bt	Rock B2	Roll B1
85 W	115 W			115 W	85 W
LHCP	LHCP	V	V	LHCP	LHCP
2333.465	2335.305	2337.490	2340.200	2342.205	2344.045

total capacity of the XM system. The exact same content is transmitted three times in three different signals for time diversity—once on each of the two satellites and a third time by terrestrial repeaters.

The downlink signals from the hub to the satellites are QPSK modulated time division multiplexed (TDM) signals with a bit rate of 3.28 Mbps and a bandwidth of 1.886 MHz. Because of forward error correction coding (FEC), the actual information rate of each TDM signal is 2.048 Mbps. The uplink frequency is X-band at 7055 MHz to 7075 MHz, and two uplink signals for each satellite are transmitted with different FEC schemes and different interleaves to achieve more robust performance against short dropouts.

The hub station uses a pair of 7 m-diam with an EIRP per carrier of 70 dBW using 3 kW klystrons. The terrestrial repeaters use vertical polarization, which expects better transmission performance under a multipath environment. The repeaters use multicarrier modulation with a rate of 5/9 FEC against frequency selective fading.

Table 4.9 shows the characteristics of the two XM satellites, which use Boeing 702 buses with payloads costumed by Alcatel. Two sets of 16 parallel 216-W traveling wave tube amplifiers (TWTA) produce 6.4 kW of transmitting power. The 5-m aperture offset-fed elliptically shaped transmit reflectors fold into three sections for launch and deploy after launch to their full size.

The terrestrial repeater, which transmits the same contents as the satellite signals to reduce signal holes in the satellite coverage, consists of a directional S-band receive antenna, an electronics box, and an S-band omnidirectional transmit antenna. The repeaters also provide some amount of building penetration. A combination of the receiving antenna's directivity, spatial isolation between transmitting and receiving antennas, and signal filtering allows the XM repeaters to receive and transmit at the same frequency band. Repeater coverage is provided in 70 major metropolitan areas, although future expansion of the repeater network is planned. User antennas have two modules—one for the satellites and the other for the repeaters.

The XM-2 spacecraft "Rock" was launched by a Sea Launch Zenit 3SL rocket from Odyssey Launch Platform floating at the equator in the Pacific Ocean east of Christmas Island on 18 March 2001, and the XM-1 "Roll" was launched by Sea Launch on 8 May 2001.

On 25 September 2001, the XM commercial services began in three cities, and in mid November 2001 the service was extended to the contiguous 48 states. The XM company announced that its subscriber numbers had reached 136,500 by the end of June 2002. The number of subscribers is expected to increase dramatically after car manufacturers begin providing satellite radios as original equipment on new car models. General Motors began installing XM satellite radio receivers in selected models, such as the Cadillac DeVille and Seville, in early 2001, and is expanding to 23 additional models in the 2003 model year.

Subscription price is $9.99 per month, or $119.40 per year (at the time of publication). Major radio manufacturers are Alpine, Pioneer, Sony, Dephi, and Audiovox. Radio starting prices, receiver starting prices, and antenna starting prices are $149, $150, and $39, respectively (at the time of publication). The number of channels is 100, including 71 music and 29 talk channels.

Table 4.9 Characteristics of audio broadcasting satellites

	Name				
	"Rock" and "Roll"	Sirius FM-1, FM-2, and FM-3	MBSAT	AfriStar	AsiaStar
Owner	XM Satellite Radio, Inc.	Sirius Satellite Radio Inc.	Mobile Broadcasting Corporation (MBCO)	WorldSpace Corp.	WorldSpace Corp.
Country	United States	United States	Japan	United Sates	United Sates
No. of satellites	2	3	1	1	1
Latitude	GEO (Rock 115°W, Roll 85°W)	47,102 km by 24,469 km	N/A	21°E	105°E
User frequency	S band (2333.465, 2335.305, 2342.205, 2344.045 MHz)	S band (2320.0–2332.5 MHz)	16 S band	L band (1467–1492 MHz)	L band (1467–1492 MHz)
Feeder-link frequency	X band (7055–7075 MHz)	N/A	25 MHz Ku band	X band (7025–7075 MHz)	
Launch vehicle	Sea Launch Zenit-3SL	Proton Block DM from Baikonur Cosmodrome	Adas 3	Ariane 44L from Kourou	Ariane 505 from Kourou
Launch date	Rock 3/18/2001 and Roll 5/8/2001	FM-1: 6/30/2000, FM-2: 9/5/2000, FM-3: 11/30/2000	4Q/03	10/28/1998	3/22/2000

(continued)

Table 4.9 Characteristics of audio broadcasting satellites (continued)

	Name				
	"Rock" and "Roll"	Sirius FM-1, FM-2, and FM-3	MBSAT	AfriStar	AsiaStar
Weight	4682 kg	2182 kg	3800 kg	2200 kg (at launch)	2777 kg
Battery power	15 kW	2.6 kW	7.4 kW	3 kW	5.6 kW
EIRP or transmit power	3 kW each transponder	N/A	2.4 kW	53 dBW	53 dBW
Bus	Boeing 702	Space Systems/ Loral 1300	Space Systems/ Loral 1300	EuroStar 2000+ / Matra Marconi Space	EuroStar 2000+ / Matra Marconi Space
Multiple Access	TDM	TDM	CDM	TDM	TDM
Channels available	100	100	50	40 radio programs and Direct Media Service	N/A
Service started	9/25/2001	2/14/2002	2004	N/A	N/A

Table 4.10 Orbital parameters of three Sirius satellites

a	Semimajor axis	42,164 km
e	Eccentricity	0.2684
i	Inclination	63.4°
ω	Argument of perigee	270°
Ω	Right ascension of ascending node	
	for FM-1	285°
	for FM-2	165°
	for FM-3	45°
	Apogee longitude	96°

B. Sirius Satellite Radio[22]

The SIRI system[22] selected unique satellite orbits that are highly inclined, elliptical, and geosynchronous with each satellite in a plane 120 deg from the others. This constellation provides high elevation angles to mobile users in the northern areas of the United States. The high elevation can reduce service outages because of blockage and shadowing from buildings and roadside trees.

The orbital parameters of three SIRI satellites are shown in Table 4.10 and their orbits are illustrated in Fig. 4.6. Each satellite spends approximately 16 h per day north of the equator, and 8 h per day south of the equator. As given in the Questions for Discussion section, the orbital period is one sidereal day of 1436 min

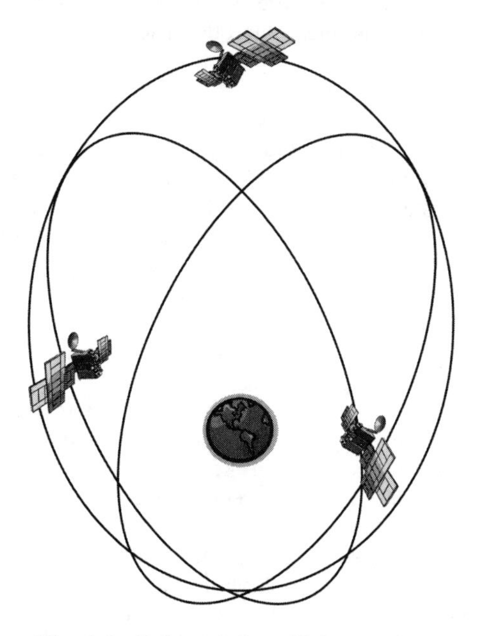

Fig. 4.6 Orbits of three Sirius satellites.

(24 h minus 4 min). When a satellite crosses near the equatorial node, the broadcasting signal from this descending satellite is switched to the ascending satellite by using onboard time tag queue commands. As the result of this constellation, one satellite is always higher than 60 deg in elevation angles to all users in, for example, the New York City area.

Three Sirius satellites, named "FM-1," "FM-2," and "FM-3," were launched on 30 June 2000, 5 September 2000, and 30 November 2000, respectively, by Proton Block DM vehicles from the Baikonur Consmodrome. A fourth satellite will remain on the ground for backup. Commercial services commenced on 14 February 2002.

Because of the unique orbit, different operations are necessary compared to the geostationary satellite systems. Because of the high inclination of 63.4 deg, the Earth station's antennas are required to continuously track the motion of satellites. The eclipse happens at different times of the year for each satellite. The Z axis of the satellites should be rotated for the solar arrays to track the sun. The payload is turned off while the satellite is in the southern hemisphere. Finally, the two satellites above the equatorial arc must operate on different frequencies to avoid interference with each other.

To enable mobile reception of high-quality digital audio signals, two TDM QPSK signals from the satellites and a coded orthogonal-frequency division multiplexed (COFDM) signal from the terrestrial antenna are delivered within 12.3 MHz band of S-band (2.3 GHz).

The subscription price is $12.95 per month, or $155.40 per year (at the time of publication). Major radio manufacturers are Clarion, Kenwood, Jensen, and Audiovox. Radio starting prices, receiver starting prices, and antenna starting prices (at the time of publication) are $180, $150, and $80, respectively. The system provides 100 channels, including 60 music and 40 talk channels.

C. Global Radio for Europe[23,24]

Global Radio is planning to provide DAB service into Europe in 2004 using Eureka-147 technology to be compatible with existing terrestrial DAB systems. For European DAB systems, selection of satellite orbits and constellations is very important for reliable broadcasting services, especially to serve high-latitude countries such as Norway and Sweden. In Ref. 23, system tradeoffs for European satellite radio broadcasting have been considered.

For European DAB services, one beam for overall European coverage and seven regional beams are required. The European beam is approximately 3.4 deg in diameter, and each of these regional beams is approximately 1.4 deg in diameter. L-band at 1467.5 MHz to 1492 MHz for satellite-DAB is used. This 24.5-MHz band is divided into 14 blocks of 1.5-MHz band and some block intervals; Global Radio plans to use nine blocks. Two of these blocks will be allocated to the pan Europe beam, one to each of the seven regional beams with frequency reuse, and two blocks to terrestrial repeaters.

In Ref. 23, they studied several types of satellite constellations such as 8-h Molinaya, standard Molinaya, left-hook Molinaya, 63-deg-inclined Tundra (Sirius), 55-deg-inclined Tundra, and GEO. Because European countries are located at a higher latitude than the United States, a different optimum orbit for Global Radio

Table 4.11 Orbital parameters of Global Radio for Europe

a	Semimajor axis	42,164 km
e	Eccentricity	0.32
i	Inclination	53–57°
ω	Argument of perigee	270°
Ω	Right ascension of ascending node	43°
	Apogee radius	55,656 km
	Perigee radius	28,672 km

will be selected. Table 4.11 shows the orbital parameters of Global Radio for Europe.

D. WorldSpace for Africa and America

The WorldSpace satellite system is delivering digital audio broadcasting services to Africa, the Middle East, and Asia by two geostationary satellites, "AfriStar" and "Asia Star," operating in L-band at 1467–1492 MHz. A third satellite needed to complete the WorldSpace constellation, "AmeriStar," is available but plans for launching it are uncertain. Each satellite can deliver approximately 40 radio programs and DMS that provides Internet content. No subscription fee is charged, as the WorldSpace business plan foresees its revenues to be obtained from advertisements on air.

The first WorldSpace satellite, "AfriStar," was launched by an Arianespace Ariane 44L rocket from Kourou, French Guiana on 28 October 1998. AfriStar is delivering 46 programs of news, music, and education in English, French, and several national languages to approximately 200,000 receivers. AfriStar has three spot beams, each of which has capacity for 96 signals at 16 kbps (mono-AM quality). For higher audio quality, several channels can be combined. The downlink signals in each spot beam are multiplexed into two TDMA QPSK signals.

"AsiaStar" was launched by an Ariane-5 rocket from Kourou on 22 March 2000. Services were started in September 2000 for the Asian areas such as Singapore, Indonesia, India, Malaysia, Thailand, and the Philippines.

E. Mobile Broadcasting Corporation in Japan

The MBCO was established on 29 May 1998 by Toshiba, Toyota, Fujitsu, Nippon TV, and Panasonic, and now has 42 Japanese companies as shareholders. The MBSAT satellite, which was designed and developed by Space Systems/Loral, will provide digital multimedia services such as CD-quality music, MPEG-4 video, and data to mobile and personal users. The MBSAT will be launched in the fourth quarter of 2003, and services will begin in early 2004.

The broadcasting signals are direct-sequence code division multiplex with a chip rate of 16.384 Mcps, operating in the S-band (2.6 GHz) with a bandwidth of 25 MHz. The MBSAT has capacity for more than 50 channels of audio and video with an S-band 12-m-aperture deployed antenna and an RF power of 2400 W produced by 16 S-band 120-W transponders. The satellite also transmits Ku-band signals to terrestrial repeaters to fill the holes of satellite coverage.

V. Conclusion

As mentioned in this chapter, many plans of mobile satellite communication systems were changed and terminated. AMSC, TMI, Iridium, Globalstar, ICO, and Orbcomm changed the initial configuration of the company, satellite constellations, and their initial service plans. Many LEO and MEO systems did not start their planned services. However, it does not mean that the MSS has failed all over the world, although the satellite phone market is much less than the predicted values. Inmarsat drastically improved global maritime communications, and today there are more than 450,000 subscribers of Inmarsat, Mobile Satellite Ventures, Omnitracs/Euteltrac, Aces, Thuraya, Iridium, and Globalstar.

Following maritime communications, aeronautical communications have to be improved by satellite communication technologies. Voice communications will be upgraded into e-mail delivery and Internet access, to broadband Internet access to entertainment in cabin on aircraft. Ku-band and higher frequencies will be used because rain attenuation will not affect signal strength on the airplane above clouds and rain.

References

[1]Jedrey T. C., (ed.), Special Issue on the Advance Communications Technology Satellite, *International Journal of Satellite Communication*, Vol. 14, No. 3, 1996.

[2]Wakana, H., Saito, H., Yamamoto, S., Ohkawa, M., Obara, N., Li, H.-B., and Tanaka, M., "COMETS Experiments for Advanced Mobile Satellite Communications and Advanced Satellite Broadcasting," *International Journal of Satellite Communication*, Vol. 18, 2000, pp. 63–85.

[3]Mastracci, C., and Cedrone, G., "ITALSAT, New Technologies Orbital Demonstration," *Proceedings of the 14th AIAA International Communication Satellite Systems Conference*, AIAA, Washington, DC, 1992, pp. 842–848.

[4]Tanaka, M., Kimura, K., Kawase, S., Wakana, H., and Iida, T., "Applications of the Figure-8 Satellite," *Proceedings of the 18th AIAA International Communications Satellite Systems Conference*, AIAA, Reston, VA, 2000, pp. 957–966; also AIAA Paper 2000-1116.

[5]Hart, N., "Mobile Satellite System Design," *Satellite Communications—Mobile and Fixed Services*, edited by M. J. Miller, B. Vucetic, and L. Berry, Kluwer Academic Publishers, 1993.

[6]Corbley, K. P., "INMARSAT-Mobile Satellite Communications from the Battlefield and Beyond," *Via Satellite*, Nov. 1995, pp. 22–28.

[7]Ohmori, S., Wakana, H., and Kawase, S., *Mobile Satellite Communications*, Artech House Publishers, 1998, pp. 292–294.

[8]Levin, L. C., and Nash, D. C., "U.S. Domestic and International Regulatory Issues," *Proceedings of the 3rd International Mobile Satellite Conference*, 1993, pp. 67–72.

[9]Pedersen, A., "MSAT Wide-Area Fleet Management: End-User Requirements and Applications," *Proceedings of the 4th International Mobile Satellite Conference*, 1995, pp. 158–163.

[10]Johanson, G. A., "MSAT Satellite/Cellular Integration and Implications for Future Systems," *Proceedings of the AIAA/ESA Workshop on International Cooperation in Satellite Communications*, 1995, pp. 155–161.
[11]Harrison, S., "MobileSat®-the World's First Domestic Land Mobile Satellite System," *Space Communications*, Vol. 13, No. 3, 1995, pp. 249–256.
[12]Wagg, M., and Jansen, M., "MobileSat® a Characteristically Australian MSS," *Proceedings of the 4th International Mobile Satellite Conference*, 1995, pp. 404–408.
[13]Warner, P., "Qualcomm Leverages Satellites for Shipping Security," *Satellite News*, Vol. 25, No. 28, 22 July 2002, p. 9.
[14]Dayarantna, L., Walshak, L., and Mahdawi, T., "ACeS Communication Payload System Overview," *Proceedings of the 19th AIAA International Communications Satellite Systems Conference*, AIAA, Reston, VA, 2000, pp. 246–250; also AIAA Paper 2000-1252.
[15]Brunt, P., "IRIDIUM-Overview and Status," *Space Communications*, Vol. 14, 1996, pp. 61–68.
[16]Hutcheson, J., and Laurin, M., "Network Flexibility of the IRIDIUM Global Mobile Satellite System," *Proceedings of the 4th International Mobile Satellite Conference*, June 1995, pp. 503–507.
[17]Hirshfield, E., "The Globalstar System: Breakthroughs in Efficiency in Microwave and Signal Processing Technology," *Space Communications*, Vol. 14, 1996, pp. 69–82.
[18]Karlin, S., "Take Off, Plug In, Dial Up," *IEEE Spectrum*, Aug. 2001, pp. 53–59.
[19]Dumont, P., Aznar-Abian, V., and Nivelle, F., "Regulatory Status on Aeronautical Mobile Satellite Service in the Ku Band," AIAA Paper 2002-1932, 2002, pp. 1–9.
[20]Michalski, R.A., "An Overview of the XM Satellite Radio System," *Proceedings of the 20th AIAA International Communications Satellite Systems Conference*, AIAA, Reston, VA, 2002; also AIAA Paper 2002-1844.
[21]Schaeffler, J., "Sirius Rolls Out; XM Roll On," *Satellite News*, Vol. 25, No. 27, 15 July 2002, pp. 4–6.
[22]Briskman, R. D., "DARS Satellite Constellation Performance," *Proceedings of the 20th AIAA International Communications Satellite Systems Conference*, AIAA Reston, VA, 2002; also AIAA Paper 2002-1844.
[23]Kidd, A., and Shomanesh, A., "Systems Tradeoffs for European Satellite Radio Broadcasting (GLOBAL RADIO)," *Proceedings of the 20th AIAA International Communications Satellite Systems Conference*, AIAA, Reston, VA, 2002, pp. 1–11; also AIAA Paper 2002-2040.
[24]Kozamernik, F., Laflin, N., O'Leary, T., "Satellite DSB Systems," *EBU Technical Review*, Jan. 2002, pp. 1–17.

Questions for Discussion

1) **Consider advantages and disadvantages of GEO, LEO, MEO, and HEO. Compare propagation delay, propagation loss, required numbers of satellites, etc.**

The advantages of GEO are configuration simplicity, extremely wide spotbeam footprint, relatively time-invariant satellite-ground terminal geometry, simple space segment control system, and fixed propagation delay. The advantages of LEO/MEO are much lower propagation delays, a much better link margin, an easier launch, and an ability to support handheld terminals. The advantages of HEO are high elevation angles even in high-latitude countries, and flexible system design.

On the other hand, disadvantages of GEO are power-limit links, excessive propagation delay for voice and ARQ-based packet data, and an inability to cover the Polar Regions. Disadvantages of LEO/MEO are a large number of satellites, more complex onboard control subsystems, less satellite dwell time, more frequent handover, and much larger Doppler shift. Disadvantages of HEO are lower link margin than LEO/MEO systems, larger onboard antenna required, large Doppler shift, and shorter lifetime because of the periodic crossing through the van Allen radiation belts.

Let us compare the propagation loss (free-space propagation loss) among these satellite systems. The propagation loss can be given by

$$L = 20 \log\left(\frac{4\pi R}{\lambda}\right)$$

where R is the distance between the satellite and the Earth station, and λ is the wavelength. For the sake of simplicity, for example, R is 36,000 km for GEO; 785 km for Iridium (LEO); 10,355 km for ICO (MEO); 26,800 km for Archimedes (HEO); and 20 km for high altitude platform systems (HAPS). Compared with GEO, $L_{\text{GEO}} = 0\,\text{dB}$, the propagation loss of these satellite systems are $L_{\text{LEO}} = -33.2\,\text{dB}$, $L_{\text{MEO}} = -10.8\,\text{dB}$, $L_{\text{HEO}} = -2.6\,\text{dB}$, and $L_{\text{HAPS}} = -65.1\,\text{dB}$.

Round-trip propagation delay between the satellite and the Earth station is given by

$$T_{\text{delay}} = \frac{2R}{c}$$

where c is the light speed, $c \cong 3 \times 10^5\,\text{km/s}$. Therefore, T_{delay} is 0.24 s for GEO, 0.005 s for Iridium, 0.07 s for ICO, 0.18 s for HEO, and 0.0001 s for HAPS.

For a satellite or a HAPS to be visible from an Earth station, its elevation angle must be larger than a minimum service elevation angle. When the central angle γ is defined as the angle measured at the center of the Earth between the satellite and the Earth station, the R_s is the distance between the satellite and the center of the Earth, and R_e is the distance between the Earth station and the center of the Earth, the center angle γ should be limited by

$$\gamma \leq \cos^{-1}\left(\frac{R_e}{R_s}\right)$$

The radius of the service area is given by γR_e.

2) **Calculate the length of the semiminor axis b, apogee altitude, and perigee altitude of the Sirius satellites using orbital parameters shown in Table 4.10. Calculate the orbital period.**

Hint:
(1)

$$b = a\sqrt{1 - e^2}$$

$$\text{apogee altitude} = a(1 + e) - R$$

$$\text{perigee altitude} = a(1 - e) - R$$

where R is the radius of the Earth, 6378.136 km.
(2) The period T is given by

$$T = 2\pi\sqrt{\frac{a^3}{\mu}},$$

where μ is the Kepler's constant, $3.9860044 \times 10^5\,\text{km}^3/\text{s}^2$.
Answer:
$b = 40{,}617\,\text{km}$, apogee altitude = 47,103 km, perigee altitude = 24,469 km
$T = 86{,}164\,\text{s} = 1436\,\text{min} = 23.9\,\text{h}$

Chapter 5

The Future of Satellite Broadcasting Systems

Edward Ashford*
SES GLOBAL, Betzdorf, Luxembourg

I. Introduction

THIS chapter examines the field of satellite broadcasting in some detail. The bulk of the topics covered concentrate on television broadcasting and distribution, but sound and data broadcasting services are also included. Likewise, while the emphasis in the chapter is on broadcasting to fixed terminals, the topic of broadcasting to mobile terminals (for example, to aeronautical, automobile, or personal terminals) is also covered. Geosynchronous Earth orbit (GEO) satellite systems are the prime focus in the chapter because the vast majority of broadcast systems are based on satellites in that orbit, but the possibilities of non-GEO systems are discussed where it is felt appropriate. The chapter has approximately a 10-year period in focus. Developments in the field that are predicted to occur beyond this planning horizon are covered in Chapter 7.

The chapter begins with a summary of the history of television standards and satellite broadcasting, describing the evolution from the early distribution of a few analog channels to the present era of digital broadcasting in which a single satellite can handle literally hundreds of channels. Various standards, both open and proprietary, used for satellite broadcasting are discussed. The subject of two-way interactive broadcast and multimedia links via satellite is explored in some detail. Terminals, both one-way and two-way, are covered, with projections indicating how these may evolve in the coming years. With this as background, the chapter goes on to explain how the introduction of digital broadcasting and broadband interactive multimedia is changing the satellite broadcasting industry and promoting convergence of these up-to-now disparate fields. This is changing conventional broadcasting into a two-way proposition, paving the way to making possible a variety of new types of satellite services. Also, the move toward various degrees of onboard processing in satellites will be highlighted because this too promises the prospect of additional new services.

The present situation of satellite broadcasting is described, as are the trends pointing toward how the field is likely to evolve in the future. Both the technology

*Vice President for Technology Development.

developments and the introduction of new types of services are discussed in some detail.

II. History of Satellite Television—The Development of Transmission Standards

Commercial television broadcasting became big business only after the introduction and broad acceptance of standards on television transmissions, which allowed broadcasters and television set manufacturers to converge on their designs. However, these standards were not universal, and different countries accepted slightly different standards. While all used interlaced images to avoid image flicker, and, eventually, a 4:3 picture aspect ratio, they differed in important details. In particular, their frame update rates were chosen to be able to be synchronized with the alternating current standard (either 50 Hz or 60 Hz) used in each region. The numbers of lines that were selected to "paint out" a picture on the TV display tube were different in different countries. Finally, the frequencies and bandwidths allocated to television broadcasters to transmit their programs differed from one country to another.

In the United States, the National Television System Committee (NTSC) agreed to a set of standards for black-and-white television in 1941. These standards were expanded and modified in 1953 to include compatible color and black-and-white transmissions, and have remained essentially unchanged since then.

In Europe, a decade later, two different standards arose. In France, a standard called SECAM (for a French acronym meaning sequential color with memory) was developed. In SECAM, red and blue color information is sent separately at two different frequencies, and stored temporarily in a memory circuit in the receiver. The color information from alternate lines is then recombined. This is done to correct the problem of sensitivity of the displayed color in the NTSC system to phase errors caused by interference and multipath effects (which gave rise to the humorous interpretation of the NTSC acronym as "never the same color").

In Germany, another system was developed that was somewhat closer to the NTSC standard, except that another way was found to avoid the NTSC color sensitivity to phase errors. In the German approach, the phase of the subcarrier containing the color information is reversed on every second line, giving rise to the acronym PAL, for phase alternate line. Both SECAM and PAL also gave rise to numerous humorous alternate definitions for their acronyms: "pictures at last," for PAL, and "system essentially contrary to America's method" for SECAM.

Of the three standards, PAL is the most prevalent worldwide today, being used in some 85 countries. SECAM comes second, with approximately 50 countries, and NTSC is a close third with 45 countries. Table 5.1 summarizes certain major characteristics of the three TV standards.

In some countries, such as Luxembourg and Monaco, both PAL and SECAM are used, which may confuse these statistics somewhat. In addition, these three standards for analog terrestrial television are not the end of the story. Depending on the country, for both PAL and SECAM, there may be differences in the

Table 5.1 Television standards

	Lines/frame	Frames/second	Nominal bandwidth (MHz)	Percentage of countries
NTSC	525	30	6	25
PAL	625	25	7 or 8	47
SECAM	625	25	8	28

frequencies at which the audio information is transmitted, how stereo sound is handled, and whether additional features such as subtitling and teletext are carried. Moreover, in Brazil for example, the color system used is based on the alternating phase concept of PAL, but with the number of lines and frame rate of the NTSC system.

When the use of satellites began in the 1960s, these three analog standards were also the basis for transporting satellite television programs. In 1964 the Olympic games were relayed from Japan to the United States through the then recently launched SYNCOM-4 satellite in GEO, and then from the United States to Europe through the inclined low-Earth-orbiting (LEO) RELAY satellite. This latter hop highlighted the difficulties in converting pictures captured and transmitted in one standard to other standards without serious degradation in image quality. Viewers in Europe apparently commented that there could be no racial distinctions in the Olympics that year—everyone's skin was a uniform shade of green!

With the advent of satellites, television broadcasting became truly international, but the nations of the world unfortunately remained divided by incompatible standards. As the decade of the 1980s drew to a close, this was still the case, although many thought that there would be a convergence in the future, as serious consideration began to be given to increasing the resolution of television broadcasts to achieve what became known as high definition television (HDTV).

A. HDTV Standards

When color television was being developed, researchers looked for, and found, approaches that would allow backward compatibility (that is, where color television programs could be broadcast that could still be seen on black and white television sets). This is true individually for each of these systems, but while the black-and-white component of SECAM transmissions can be viewed on a PAL receiver and vice versa, and with some minor changes they can also be seen on an NTSC set, the color portions of the three types of systems are not compatible.

Beginning as early as the late 1960s, people around the world began to experiment with systems to improve the picture quality of television images. Initially, methods that would allow backward compatibility were sought, as in the case of color and black-and-white compatibility. In most cases, however, this goal was found to be too difficult, and HDTV systems, having more scan lines than used previously, began to be pursued as totally separate objectives. Elsewhere, albeit somewhat later, researchers developed systems with a similar increase in the number of horizontal scan lines, but with a wider screen format ratio of 16 : 9, rather than the 4 : 3 ratio of standard definition television. The HDTV systems investigated all had more than five times the visual information provided in an NTSC broadcast. This greatly improved the realism of the programming, because it gave the viewers the illusion that they were somehow part of the scenes they were watching, rather than outside observers.

1. The Japanese Standard

In 1969, researchers at NHK in Japan began developing a high-definition system, combining analog and digital techniques, which would eventually result in what became known as the multiple sub-nyquist encoding (MUSE) system. This system, and its improvements over time, had 1125 lines, with interlacing based on the 60-Hz standard, and used a picture tube with an aspect ratio of 5 : 3 (that is, 15 : 9 instead of the 16 : 9 used elsewhere in HDTV experiments).

Improved quality generally came at a high price, however—much more bandwidth was required. Experiments conducted starting in 1978 in Japan, to send HDTV transmissions via satellite, required some 100 MHz of bandwidth, or the equivalent capacity of almost three satellite transponders in parallel. Given that satellites of that period typically had only 10–12 transponders onboard in total, it would have been rather expensive to pursue the provision of such satellite HDTV programming on a commercial basis. The NHK researchers continued working, however, and eventually developed a more efficient version of MUSE that would fit within a single satellite transponder. That MUSE system was then put into operational use in Japan.

2. The ATSC Standard

In the mid-1980s a number of U.S. companies began developing systems that they hoped would be competitive with the MUSE system. Unfortunately, all of the manufacturers decided to go their own way, and as many as twenty different types of enhanced definition or high-definition systems were pursued, some using analog technology and others using digital. The Advanced Television Systems Committee (ATSC) was formed in 1983 to help resolve the differences between these systems and to implement an Advanced Television Standard for the United States.[1] In 1987, the Federal Communications Commission (FCC), in an attempt to resolve this situation, had the systems tested to determine the best and, by early 1993, had narrowed the field down to four competing systems from three groups of companies. These companies were pushed by the FCC to find a compromise between their competing designs, and in May of that year they agreed to form a Digital HDTV Grand Alliance that would combine the best aspects of each of their systems. They further agreed to share royalties on the marketing of the finally selected single system.

The Grand Alliance eventually settled on a single, digital standard, which was later adopted by the ATSC, and became known as the ATSC system. This system uses MPEG-2 coding for video, compressed audio according to the Dolby AC-3 standard that allows multichannel audio to be transmitted in parallel, multistage forward-error correction, and vestigial-sideband (VSB) modulation for terrestrial broadcasts [quadrature phase shift keying (QPSK) is in general use for satellite transmissions]. The ATSC standard is not restricted to HDTV, and can equally be used for the transmission of standard definition television. When used for HDTV, however, the format specified is 1080 lines in a 16 : 9 aspect ratio with a 30 frame/s interlaced scan. The first satellite transmissions using the new standard started in 1994, using a "condominium satellite" that carried transponders for two U.S. satellite broadcasters, DirecTV and USSB.

3. *The DVB Standard*

Researchers in Europe were also busy while these activities were going on in Japan and the United States. Initially, in the latter part of the 1980s, the concentration was on improving the quality of analog transmissions in a manner similar to the MUSE system in Japan, using analog-digital, sub-nyquist algorithms. The system was named MAC for multiple analog components. Several approaches were investigated, including C-MAC, D-MAC, and D2-MAC, finally leading to an HDTV standard known as HD-MAC. This did give considerably improved quality video and sound, but suffered from the amount of computation required at the transmitting end to encode signals in the standard and, more importantly, the complexity and the sensitivity of the receivers necessary to decode it. Nevertheless, Europe experimented with satellite broadcasts using this standard, and it came close to being accepted as a general standard until parallel developments in digital television were demonstrated that could give an equal-quality picture with less bandwidth.

In parallel to the efforts to obtain a consensus on the hybrid analog/digital MAC standards for satellites, researchers in Europe were also investigating full digital implementations for terrestrial HDTV. In Germany, a group of researchers was formed to examine the various types of digital systems being developed elsewhere in the world, with the goal of being able to select the "best of the best" from these for Europe. This group was soon joined by other members from outside Germany, and became known as the European Launching Group. This group continued to evolve, and in 1993 it metamorphosed into the European Digital Video Broadcast (DVB) Project group. As of the end of 2001, the group had more than 270 members from some 30 different countries, representing content providers, hardware manufacturers, network operators, European international institutions, satellite operators, and governmental regulatory bodies.

Although the original European focus concerned the problem of digital standards for terrestrial HDTV, it was soon realized that the agreement on a digital standard was what was most important. The emphasis on HDTV was reduced when it was realized that, provided a digital standard was flexible enough, it could be compatible with a wide range of media types, of which HDTV was only one. Moreover, it could cater to any of the various television standards then in use, making it a global standard potentially.

In late 1993, the first European digital satellite transmission standard was approved by the DVB Project Steering Committee and sent to the European Telecommunications Standards Institute (ETSI) for its consideration. Finally, in November 1994, ETSI approved and released the DVB satellite specification as an official European standard.

However, this was not the end of the story. The DVB-S standard for satellites is aptly suited for the satellite environment, but is not matched to terrestrial broadcasting or cable usage. The DVB project, therefore, has continued to remain in existence, and has produced a family of standards for DVB transmissions, including those optimized for cable (DVB-C), terrestrial (DVB-T) and microwave (DVB-MC and DVB-MS) transmissions.

B. Standards?

You may be asking yourself, 'If each of the cases mentioned (and there are even more that have not been mentioned) has a different specification, what is meant by the word "standard" in the terms "ATSC or DVB Standard?" ' The simple answer is that the different standards are in fact each a portfolio of substandards and specifications that can be selected, according to certain rules, defining a common approach that all systems adhering to the standards must take. Moreover, for the most part, there are great similarities between most of the digital satellite broadcast standards in use today, with the differences between them often being because of rather narrow details. In what follows, therefore, the discussion will concentrate on the system in use in Europe for digital television, the Digital Video Broadcast (DVB) standards, because this system is now the most commonly encountered internationally. Certain differences between the DVB standards and the ATSC standards in use in the United States will be pointed out in the text.

The rules for DVB television transmissions are, for example, that they be

1) Compressed, using MPEG-2 compression for video and MUSICAM (masking-pattern universal sub-band integrated coding and multiplexing) compression for audio (as previously mentioned, ATSC uses another form of audio compression developed in the U.S. by Dolby). Such compression, which results in little discernible degradation in quality, provides the possibility of multiplexing a number of television programs into the bandwidth that would be required by a single program in an analog system.
2) Combined with service information that the receiver uses, for example, to extract specific programs and to construct an electronic program guide.
3) Provided with the means to ensure conditional access (granting access only to selected groups of users) through scrambling/encryption.
4) Coded with a Reed–Solomon error-correcting outer code to allow automatic recovery of bit errors or lost bytes.
5) Packetized into fixed length (188 byte) MPEG-2 frames and efficiently modulated, where the type of modulation may be selected according to the transmission means envisaged.

Various tools exist to supplement these rules, to make it possible to provide an optimized "tailor-made" standard for specific cases. For satellite broadcasts, for example, additional convolutional interleaving and coding is used to protect against the bursty types of errors that such systems are prone to. Because power is always at an expensive premium on a satellite, QPSK modulation is typically selected for satellite broadcasts, because it is one of the most power-efficient forms of modulation (a variant of QPSK, "Offset-QPSK" or "O-QPSK" is used in the U.S. on the Hughes DirecTV DSS and other systems).[2] For terrestrial broadcasts, where power is a much less important parameter, more frequency-efficient means of modulation can be and usually are selected.

Digital satellite television is essentially a "one-to-many" type of service. A number of different television programs are aggregated, digitized, compressed, multiplexed together according to the appropriate set of standards, and uplinked to a satellite as a bouquet of programs in a single TDM (time division multiplexed) bit stream. The satellite then downlinks this TDM stream down to many users in a wide area of coverage.

At the end of 2001, nearly 18 million households were able to receive "direct to home" (DTH) satellite television in the U.S., with almost all of this transmitted in digital form. At the same time, there were some 38 million households receiving DTH satellite television broadcasts in Europe (plus an additional 64 million receiving cable television that was fed by satellite). Of the 38 million, the number receiving digital TV approached 17 million, with the rest still receiving analog transmissions. However, the percentage of DTH customers receiving digital transmissions has been increasing steadily and, by the end of 2002, there were more digital satellite receivers in Europe than analog. Digital satellite television is, therefore, a rapidly expanding market.[3]

The preceding technical description is valid for television, but the digital satellite broadcasting standards are also applicable to IP traffic, asynchronous transfer mode (ATM) data, and many other types of multimedia transmissions. With DVB, for example, such traffic, which may or may not be first compressed, is encapsulated into the DVB format and into the fixed-length MPEG-2 frames. These frames are then multiplexed into a TDM data stream together with other such frames or with frames carrying television. The multiplexed data stream is then transmitted via satellite.

So far, only the forward-link standards have been discussed, which were initially developed for one-way broadcast of television. With the explosive rise of the Internet, however, satellites have begun to play an increasingly important role for broadcasting data. Information can be broadcast over a satellite to PC users at data rates far exceeding the rates possible over public switched telephone network (PSTN) lines, and over areas well beyond the reach of either television cable or asynchronous digital subscriber lines (ADSL). A user sitting at his PC, however, needs some way to communicate to his ISP to let it know that he wants a specific bit of information. This can and is often done via a PSTN connection, but it was recognized that a return link via satellite could provide far more speed and flexibility than the use of the PSTN. This recognition gave rise to the need to develop a return channel DVB standard. In 1999 a subgroup of the DVB Project was formed to address this need. The ad-hoc DVB-RCS (return channel system) group tracked developments by key satellite operators and followed a number of pilot projects organized by the European Space Agency (ESA). That activity led to the ETSI approval, in 2001, of the DVB-RCS specifications governing the return link.

The DVB-RCS specification calls for a satellite terminal [RCST, sometimes known as a satellite interactive terminal (SIT) or satellite user terminal (SUT)] supporting a two-way DVB satellite system. The SIT receives a standard DVB-S transmission generated by a satellite hub station. Packet data may be sent over this forward link in the usual way. The same antenna is used for transmission in the return link. Transmissions use a multifrequency time division multiple access (MF-TDMA) scheme to share the capacity available. The return channel is coded and the data transported either encapsulated in ATM cells or using a native IP encapsulation over MPEG-2 transport. It also includes a number of security mechanisms. All transmissions by a user terminal are controlled by a network control center (NCC). Before a terminal can send data, it must first join the network by logging on with the NCC using a separate frequency shared between all user terminals.

The DVB Project has been very prolific. At the time of publication, some 33 different specifications had been approved and had become European standards for satellite and terrestrial communications, together with some 19 guidelines for their implementation. This is, however, not the end of the story. By the end of 2001, the DVB Project had essentially completed three of the four phases of activities envisaged for it. The fourth will cover further expansions of the standards to cater for additional Internet transport mechanisms, including, in particular, interoperability with future mobile networks. This may bring DVB full circle to reconnect with the other European standard that has spread to most parts of the world, the GSM mobile phone standard. It is difficult to predict the future of DVB. One thing is certain, however, that regardless of where they were established, the DVB standards are not the property of any one region. They are truly global and available for all to benefit from.

That leads to a final point. This section has discussed the European DVB Standards, but in fact they are far more than that. In one form or another, they have been accepted in wide areas of the world, and are under consideration in many more. Because they are open standards, and because of their inherent capability and flexibility, these standards have been recognized as a positive step forward wherever they have been seriously examined. Perhaps in a few years' time they may become the first truly worldwide set of such standards. I am sure that manufacturers around the world would welcome the extent of the market expansion in the field that this would bring.

C. What Comes Next?

Up to this point, this chapter has addressed events in the past, leading up to the situation in the world today, where there now exist highly flexible digital satellite transmission standards and many millions of homes with satellite TV receivers and dishes making use of them. From here on, the chapter will shift mental gears a bit and discuss technology advances that, more and more, will allow services other than television to take full advantage of the digital broadcast standards.

III. Satellite Broadcasting Trends at the Beginning of the 21st Century

A. Industry Trends

Satellite operators have been doing rather well financially. The top ten operators worldwide had combined revenues in 2002 of more than $5.4 billion.[4] Of the various types of services these operators provide, satellite television broadcasting, either direct-to-home or to cable heads, is by far the largest money earner. Moreover, it is expected to remain the dominant service throughout at least the first decade of the 21st century.[5] There are, however, trends that are exerting pressures on satellite operators, which are in turn putting pressure on all elements of the broadcast value chain, from satellite manufacturers through to content providers and broadcasters. Some markets are showing signs of becoming saturated. Television is a media enjoyed by practically every household in the developed world. While television started with terrestrial broadcasting, in many places this has become for the most part (and certainly in terms of how many channels can be received) supplanted by either cable or satellite television. There are, however, only so many households in any region that can and will subscribe to these TV services, and when all of these are already subscribers, there is little room for further growth, at least in the number of receivers. In Western Europe, for example, there are already signs that this saturation point may not be far away. Other factors also arising are

1) Consolidation of broadcasters through mergers, and broadcaster attrition through bankruptcies are reducing the demand for satellite capacity.
2) Improvements in data compression technology allow more channels to be put through each transponder, also depressing the demand for additional capacity.
3) Full transponder lease durations of 10 years or more are giving way to shorter-duration leases of fractional portions of the capacity of a single transponder. This then is increasing the marketing costs of satellite providers.

These trends are all acting to reduce the demand for leases of satellite transponders for television, which has been the "bread and butter" source of revenue for satellite operators for many years. Is there no hope then for the future of satellite operators? Of course there is. Geographical expansion is one key to success. There are many geographical regions that are, and will continue to be, underserved by television broadcasters, and expansion into these areas will allow operators to continue to increase their revenue stream growth for many years to come. Many operators are, themselves, consolidating and merging with others, increasing both their geographical reach and affecting improvements in operating efficiency.

The second key to future success for operators is to move into additional or new markets. Provision of broadband Internet and multimedia services, either to connect ISPs to an Internet backbone or to allow individual users broadband access to their ISPs, are services that are growing steadily. This, together with the new services that will arise after such connectivity is available to a wide audience, will allow operators to tap new sources of revenue. Some predict these sources will eventually outstrip the revenues enjoyed by operators today in the provision of "traditional" DTH services (if a period of only a couple of decades or so of operation can be considered traditional). Furthermore, the technologies that will be needed, both in space and on the ground, to efficiently serve this new type of market are being developed today, and should soon enter the marketplace.

B. Technology Trends

1. GEO Orbit Crowding

As indicated in Chapter 2, the geostationary orbit, in particular, is a finite resource with only a limited number of positions at which satellites can be placed without interfering with other satellites. As more and more GEO satellites have been put into operation, satellites have been forced to occupy positions closer and closer to one another. The exact number of satellites actually in GEO is not clear, because different references quote significantly different values. The online subscription reference resource, "The Satellite Encyclopedia—TSE"* lists some 280 satellites in the geostationary orbit, while Ref. 6 estimates the number as being much higher, with perhaps as many as 550–570. Whatever the actual number, however—and more are being launched each month—while there is little likelihood of orbital crowding of GEO satellites actually resulting in collisions, at least in the foreseeable future, the probability of satellites interfering with each other's operating frequencies is another matter.

To avoid such interference, satellites in the future will have to use higher-frequency bands. C-Band and Ku-Band predominate now, but Ka-band will soon come into its own, both for broadcasting and for two-way multimedia interconnectivity.

2. Increasing Demands for More Bandwidth

As more and more Internet users begin to experience the advantages of a broadband connection, users who rely on satellites for their Internet access will begin to demand broader-band connections. Many of the new services that will be discussed later in this chapter will also increase the demand for satellite bandwidth considerably.

There is an old adage to the effect that the safest investment one could make is in real estate. The same thing could equally be said for the frequency spectrum. There is only so much, and as demand increases, there is only one direction to go—up in frequency. So, as in the case of the Clarke orbit overcrowding, the relentless tendency in the future will be to migrate new services to higher frequency bands. Initially, these will use the Ka-band frequencies, but before too many years have passed, the use of even higher frequency bands will become prevalent.

3. Blurring of the Distinction Between TV and PC

It used to be easy to distinguish between the types of services offered via television and those available to PC users. Television viewers watched movies, soap operas, and news broadcasts, and PC users downloaded files, sent and received e-mails, and surfed the Web for information contained on an almost unlimited number of Web pages "out in cyberspace." There was an equally wide distinction between the users of each type of media: PC users were "techno-geeks" or nerds, and TV viewers were "couch potatoes."

---

*Data available online at http://www.tbs-satellite.com [cited March 2003].

These distinctions have become blurred over the years, until now there are no black-and-white differences between them, only varying shades of grey. In Europe, in particular, the first breach in the digital divide between the two types of media was probably the growth of teletext services. These made use of the fact that, in an analog TV program, there are recurrent times during which no picture information is being transmitted. These are the times in an interlaced TV transmission when the electron beam (in a conventional cathode ray TV display) is moving from the bottom right-hand corner of the screen, having completed one subframe, to the top left-hand corner of the screen, to begin the next. This "lost" time can be used to transmit, over time, many "pages" of data that can, for example, be stored in memory in the TV set, to be displayed when called up by the viewer using his remote control. The second breach in the divide was when com-panies developed software and compression algorithms that allowed "streaming video" to be received on a PC monitor, over relatively low bandwidth dial-up networks.

The third breach, and perhaps the one that will eventually break down all the remaining apparent barriers between the two types of media, was the advent of digital television, and particularly that of digital satellite television. A TDM bouquet of TV programs need not actually carry televison at all. The encapsulated data can be MPEG compressed televison, computer data files, CD-quality audio broadcasts, Internet Web pages, or a combination of all of these. Digital is digital, it is said, and "all bits are created equal." Because the digital bit streams do not distinguish between different types of transmissions, why should the user? The only real reasons that there is still a significant difference in the two types of media today, are a social one (nerd vs couch potato), and a technical one (TV screens do not have the resolution and clarity of a PC monitor, and hence are not good at displaying, for example, small letter fonts in a text document). Both of these reasons are in the process of disappearing.

4. *Television Interactivity*

The Internet is a two-way entity. While the data flow in each direction is not symmetric, a user sitting at his PC generally must initiate a command to download a Web page, a data file, or a music file. Television, on the contrary, has always been a one-way medium. All the channels available are broadcast at once, and a viewer simply tunes in to watch what is wanted. Even when a viewer uses his remote control to select a specific program from the electronic program guide (EPG) that is broadcast together with his received digital bouquet of programs, he is really interacting only with his set-top box, not with the broadcaster.

Things are, however, beginning to change. Interactive satellite and/or cable television is already available in many places, and this service is growing. While this is based on using a PSTN or ISDN line as the interactive return line today, developments are underway to use satellites both ways, liberating users completely from reliance on terrestrial communications links.

C. Communications Technology Developments

To cope with the challenges put forward by the trends previously mentioned, a number of improvements and/or breakthroughs in communications technology will be needed. (Improvements in bus technology are further covered in Chapter 7.)

1. Multibeam Satellites

To be able to reuse the frequency spectrum multiple times, a satellite will have to be able to generate many small circular spot beams covering the region in which they are operating. For coverage of Western Europe at Ka-band, for example, as many as 48 closely spaced spot beams could be used. To cover the continental United States, roughly 65 beams could be used. If there are less than this, there would be a reduction in the amount of frequency reuse that would be possible. If there are more than this, the size of the beams begins to become less than approximately 0.5 deg, so the present limits on how well the attitude of a satellite can be controlled to maintain its antenna beams pointing in the proper direction (roughly $+/-0.05-0.10$) begin to become the predominate error source, tending to actually reduce rather than increase the actual throughput capacity of the satellite.

With multiple spot beams, it is possible to use the same frequency spectrum in more than one beam. As long as the two beams are far enough apart (typically, this means being separated by at least one antenna 3 dB beamwidth), the same spectrum can be used in each without significant interference. For the size beams previously mentioned, for either European or U.S. coverage, the same frequencies could be reused essentially every four beams. Thus, for the 48-beam European case, the available spectrum could effectively be increased by a factor of 12, and up to a maximum of 16 for the U.S. case. This then can increase the throughput capability of a satellite by an equal factor. Satellites with throughputs of 10 Gbits/s or more are under development today, and as will be shown in Chapter 7, capacities in the terabites/second range could be available in the foreseeable future.

To form such a large number of narrow, high-power spot beams, however, is not an easy task. They can either be formed by a conventional multiple feed set, with a conventional reflector, or by a phased-array system (albeit at very low power efficiencies with present technology). In either case, however, the size of satellite required and the corresponding solar array power level are at the upper ends of the state of the art in bus technology.

Advances in phased-array antenna technology in the future may make it possible to overcome the "0.5 degree attitude control barrier" previously mentioned. Phased arrays can be designed to be able to dynamically repoint beams in real time. While present-day limits on the accuracy with which a satellite can be pointed put a lower limit on the size of spot beam that can be considered, the limit on the short-term stability with which a satellite can be pointed is perhaps an order of magnitude better than its pointing accuracy. Thus, if the beam pattern of a multibeam satellite can be measured on the ground and fed back to the satellite in real time, the beam control parameter of its phased array antenna could be adjusted to compensate for initial pointing inaccuracies.

The primary theoretical limit to such an approach lies with the terms "short-term" and "real time" used in the previous paragraph. Any changes with time in the satellite's attitude must be negligible compared to the finite time it takes for its transmissions to reach the ground, be measured, and point information fed back to the satellite. Even advances in technology cannot overcome the ultimate speed of light limitation.

2. Higher Order and Adaptive Modulation and Coding

The theoretical spectrum usage efficiency of BPSK modulation is 1 bps/Hz. QPSK, which is commonly used in satellite communications links because of its robustness in the presence of noise and its power efficiency, can achieve a theoretical spectral efficiency of 2 bps/Hz. Cable television distribution systems, which can benefit from the lower noise environment in fiber cables compared to satellites, and which are not so power constrained as are satellites, may use 64QAM, 128QAM, or even 256QAM, achieving theoretical efficiencies of up to 8 bps/Hz.

A primary factor limiting the modulation order used with satellites is that of onboard power. Sufficient power capability must be available on the satellite to get its signals through in the face of both noise and of signal attenuation over the long path length between satellite and receive terminals on the ground. This attenuation, moreover, is not only because of the distances involved but also because clouds and rainstorms on Earth themselves add additional attenuation to an already very weak satellite signal. At Ku-band, such weather attenuation can typically go up to 6 dB or more, and at Ka-band, it can be more than twice this. This attenuation must be considered in the overall design of satellite systems, and results in it being necessary to include rather high margins in satellite power.

Bad weather conditions do not occur all the time, of course, but users will reject a system that has frequent outages. The concept of availability (in simplest terms, the percentage of time during which a system is fully operational) is an important one where the acceptance by the general public is concerned. Even a system that is available 99.9 percent of the time will have outages during one year of almost nine hours. If these occur in the middle of a Super Bowl game or a World Cup match, there would be a (satellite television viewing) public outcry, to say the least! Hence, the need for margins on board the satellite.

Up to now, the power margins on broadcasting satellites have been essentially wasted during those times, which are most of the time in fact, when there has been negligible weather attenuation in the satellite transmission. Why not use the margins to provide the excess power needed to use higher-order modulation than the conventional QPSK? When there is little or no weather attenuation, 8PSK, 16QAM, or even higher-order modulation could be used if there was sufficient power to keep the received signal-to-noise ratio sufficiently high.

Of course, one cannot broadcast a television program at normal speeds, using 16QAM for example, when there are clear weather conditions, and then have to slow it down to lower frame rates when bad weather is encountered. There is a minimum bit rate that must be guaranteed for television or video transmissions. For data, however, this is normally not the case. As long as the link connection is not broken entirely, most users would be happy to accept broadband connection speeds most of the time, with a reduction in connection speed during those short and relatively infrequent times when there is particularly bad local weather.

Using high power is not the only way to guarantee that a signal will get through in times of high attenuation. One can use more effective types of signal coding when the quality of the link is poor, to accomplish the same objective. The overhead of using high levels of coding on signals reduces the rate at which the signals themselves can be sent, of course, but for data, the situation is similar to that previously described for adaptive modulation. As long as the link is not broken, users typically remain happy. These then are the basic ideas behind adaptive modulation and coding. Techniques and modem hardware and software are under development that will make use of both of these in the not too distant future.

3. *Onboard Processing*

Broadcasting satellites of today, with few exceptions, use bent-pipe transponders to relay their programming from an uplink station, via the satellite, to receivers on Earth. This is perfectly acceptable for television transmissions, as well as for the majority of data applications. There are, however, certain applications where bent-pipe transponders may not be acceptable. When users wish to use satellites for making telephone calls (or video-telephone, which may predominate in the future) directly to other users in a multibeam satellite environment, problems arise with bent-pipe transponders. If a user is in one beam, and the intended recipient of the call is in another, there must be an efficient means of routing the call to the correct beam on the satellite.

There are several types of system architecture that can be considered to do this. Conventional very small apaerture terminals (VSAT) have been providing direct terminal-to-terminal connection possibilities for many years, but the efficiency of such links has been poor. Only a relatively limited number of VSATs can use a satellite transponder simultaneously, and then usually only at rather modest data rates.

A second type of architecture using bent-pipe transponders involves a user uplinking his digital signal (having a header that identifies the terminal for which it is intended) to the satellite (using CDMA or MF-TDMA, for example, to allow simultaneous or near-simultaneous use of a transponder by many users). The satellite then sends the aggregate of all the signals received on that transponder to a hub station on the ground. In the hub, the signals are demodulated, decoded, and the header is read to identify the intended recipient of each transmission. The signals to any specific satellite downlink beam are then aggregated together, recoded, remodulated, and sent up to the satellite using the appropriate uplink frequency band corresponding to the downlink beam in question. The intended recipient then receives the message in his beam.

This process, however, involves a double hop, with signals having to be sent to and from the satellite twice, with each hop involving at least a one-quarter second delay because of the finite speed of light. The double hop, therefore, takes at least one-half second, plus whatever time is needed at the hub to process the incoming and outgoing signals. If one is exchanging files or sending streaming video or web-cam signals, this time delay is not normally even noticeable. If there is real-time interaction, however, such as in a telephone conversation or a video conference session, this is not the case. While the quarter-second delay involved in a single hop through a satellite is tolerable to users, if even noticed at all, the same is not true for twice this delay. Most people do not have the self-discipline to use the “over” verbal command technique (meaning “end of transmission, and awaiting your response”), which should be familiar to people having used simplex communications links in the military or as ham radio operators, or having watched movies where this was done. Most people tend to begin talking before someone on the other end of the phone even stops talking. With negligible time delay, this either results in the party on the other end breaking off his speaking, either through courtesy or to hear what is being said to him, or both parties being stubborn and continuing to talk, which is irritating to both parties. When there is a half-second latency delay, the latter situation is the one that prevails, and users on both ends of the link invariably become irritated.

To avoid this problem, one can move the processing function that was previously described as taking place in the hub station into the satellite itself. With such an onboard processing (OBP) payload, the need for the link down to the hub and back again to the satellite is avoided, and only a single hop is necessary. Latency reduction is, however, only one of a number of benefits that OBP payloads can provide. For example, by avoiding the loop through the hub, the frequency bands that would have to be used for that are kept free for other uses, thereby conserving valuable frequency spectrum. Moreover, an OBP system can be designed that can not only package messages into the correct spot beam, but also "store" messages destined for a particular downlink until a sufficiently large number are present to make the most efficient use of that beam. This improves the overall throughput of the satellite. The duration of such storage must, of course, be limited to only a few tens of milliseconds; otherwise, a latency problem like that of a double hop could occur.

It is worth noting that the problems that latency can cause (as well as problems that it was thought that latency could cause, albeit incorrectly) were one reason put forward in support of the development of several LEO communications satellite systems in the 1990s. However, LEO systems are not necessarily exempt from latency problems themselves, as a hypothetical example will show.

Consider the case of a constellation of LEO satellites, such as was originally proposed for the Teledesic system, in which there were to be 21 orbital planes, each having 40 active satellites orbiting at a maximum of 705 km altitude. Let us consider the case of just one of these planes, and suppose that each satellite is interconnected to the satellite in front and behind it by intersatellite links (see Fig. 5.1). In this plane, the intersatellite orbital spacing is 9 deg, and the distance between each satellite is 1111 km. Consider two Earth stations, located in the same plane as these satellites, separated by a subtended angle θ. It is easy to show that, for such a geometry, the minimum number of satellites through which a signal must pass in going between the two Earth stations is given by

$$n = \frac{\theta}{9}, \tag{5.1}$$

and the total distance that the signal must travel is

$$d = 1762 + 1111(n - 1) \tag{5.2}$$

Suppose we have two stations 180 deg apart, therefore requiring a total of 20 satellites. The distance the signal will have to travel is 22,871 km. This distance will introduce a small amount of latency, but there is also another source that actually can be much larger, as we will see.

Satellites using intersatellite links must have a degree of onboard processing in their payloads, if only to determine whether to send a signal onward to the next satellite or to send it down to the next Earth station. This processing will introduce a finite delay between receiving a signal and retransmitting. This delay will be even greater if a regenerative OBP is used, because of the amount of computation that

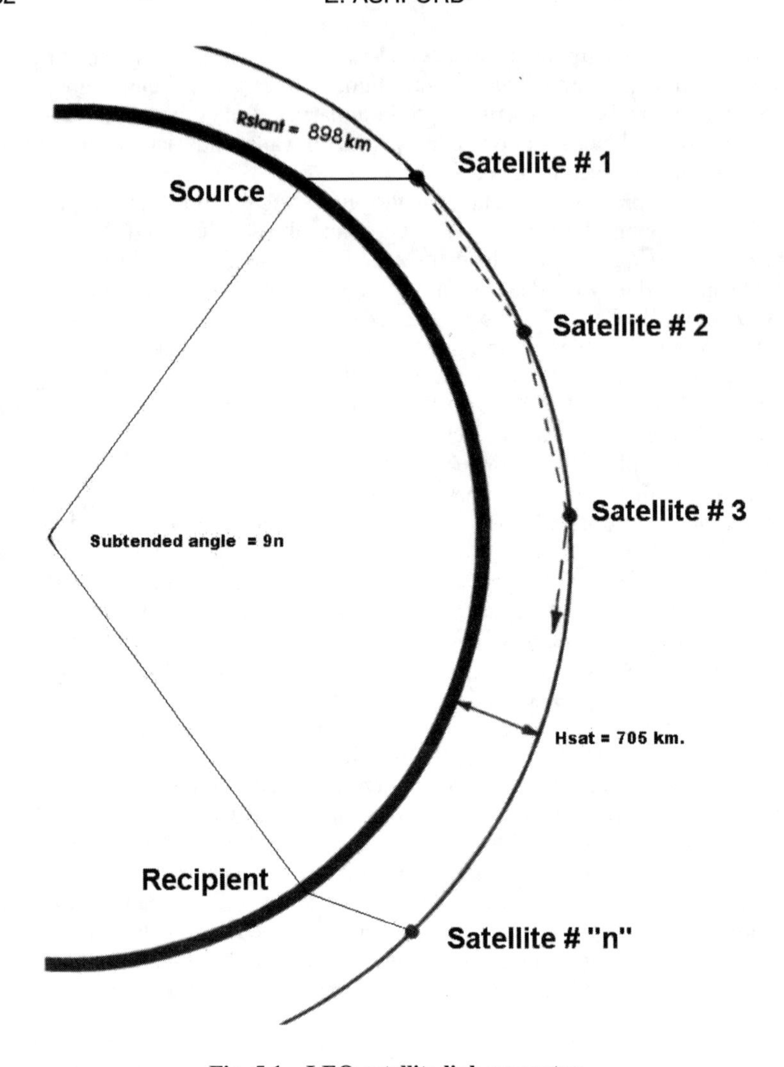

Fig. 5.1 LEO satellite link geometry.

must be done to demodulate, decode, correct errors, route, recode, and remodulate the signals. While modern OBP payloads are very fast, the total delay in transiting through a satellite can be of the order of some milliseconds. Let us call the delay encountered in each satellite τ. Then the total latency T_L in the signal between the two stations we have considered will be

$$T_L = \frac{d}{C} + n\tau$$

$$= \frac{[1762 + 1111(n-1)]}{C} + n\tau$$

$$= 76.3 + 20\tau \text{ (ms)}$$

We see that it is necessary to keep the processing delay as small as possible, because even if it is only 10 ms, the total latency will exceed that encountered in a GEO link.

In a bent-pipe repeater system, any noise on the uplink is amplified onboard the satellite and transmitted on the downlink, where it adds to the noise contributions on the downlink. In an OBP payload, the forward error correction attached to the uplink signals can be used to correct most, if not all, of the errors encountered on the uplink. Such a regenerative repeater can, therefore, greatly enhance the quality of transmissions.

Finally, an OBP payload can allow a multibeam system to be used in a broadcast mode more efficiently than otherwise. Generally, multibeam systems are not considered efficient for broadcasting in that using them for this purpose defeats one of the prime advantages of such a system, namely multiple frequency reuse. Where broadcasting must be done to only a limited number of beams, however, an OBP system can use bit replication onboard the satellite to send the same uplink message to several downlinks, thereby avoiding the need for multiple uplinks of the same content to achieve this.

4. *Cheap Terminals*

The final area of technology I will explore, for which extensive research and development is underway, is that needed to be able to produce inexpensive two-way (relatively) broadband satellite access terminals. The primary reason that DTH satellite televison is "big business" today is that affordable television receive terminals have been developed. The complete terminal, consisting of the outdoor unit, indoor set-top box, and interconnecting cabling can now be obtained for less than $200, or even less if a degree of subsidizing of the costs is involved, as it often is. The same is not true today, however, for two-way, wide-band terminals. The DirecWay® and StarBand® in the U.S. may be approaching affordable levels, but are thought to be heavily subsidized (which may be one reason that StarBand filed for bankruptcy in 2002). These terminals are, however, only really medium-band terminals. What is needed are terminals that can support tens of megabits per second in the forward link, and up to several megabits/second in the return direction, at unsubsidized prices in the range of $300–$400 or less. When this is possible, the broadband satellite business can really be expected to take off.

IV. New Services in the Coming Decade

Having described the technologies that we can expect to see in the coming decade, the question arises, "What new types of services will arise to make use of these technologies?" The following sections of this chapter list a number of these new services.

A. Internet Access via Satellite

Connections to the Internet are inherently asymmetric in nature. A web-surfer usually wants to download files or web pages having far more content than the short messages he or she sends upstream to ask for them. An individual user can therefore connect to the Internet via an ISP (or other sources of information or entertainment content) using a narrowband PSTN line for example, and have the files or Web page sent via one of the broadband TDM satellite downlinks. In the United States, Hughes Network Systems has offered such a service for several years on their DirecPC® system, targeted at residential users, that can download files at up to 400 Kbps.[1] In Europe, SES ASTRA has offered a similar service, called AstraNet®, for several years. AstraNet can deliver files to users at speeds up to 38 Mbps, far exceeding anything that has commonly been available through terrestrial means. Figure 5.2 illustrates the basic principle of operation of such one-way via satellite Internet access systems.

The process of aggregation used for digital satellite television, where multiple programs can be combined into one TDM signal stream, can also be used for data. Multiple data streams from many different sources, which could be from the Internet or from any other source of multimedia content, can be aggregated together at a gateway site and uplinked in a TDM stream that can be received by anyone in the antenna coverage area of the satellite's downlink.

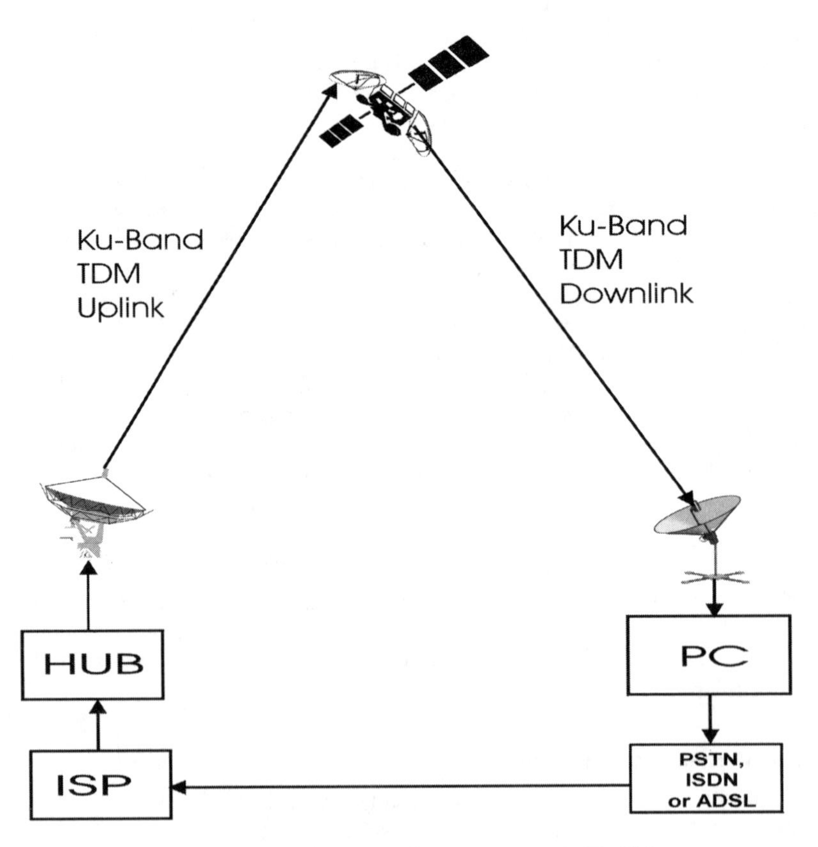

Fig. 5.2 Internet access via a one-way satellite link.

One-to-many services make the best use of satellite resources. In these services, the number of people receiving information in the downlink can grow to any size without saturating the satellite capacity. In the business-to-business category, companies in Europe are beginning to make use of this to propagate company databases, marketing or training videos, for example, from their headquarters to many geographically scattered divisions or subsidiaries.

One way to implement such one-to-many business services, or even one-to-one services, is for a company to procure a two-way satellite terminal, from, for example, (in the United States) Hughes Network Systems (a DirecWay terminal) or from StarBand (Skyblaster® terminal) or (in Europe) an SES Broadband Interactive (BBI) terminal, and locate it at their headquarters site.

With a BBI DVB-RCS multifrequency TDMA terminal, for example, companies can uplink a data stream via satellite of up to 2 Mbps at Ka-band to the SES ASTRA-1H satellite, containing whatever data they want to send to their outlying plants. From the satellite it will be downlinked to a BBI hub. The hub turns the stream around, multiplexes it together with other data streams, and sends it in a Ku-band DVB-S TDM stream back up to either the same satellite that originally handled it, or to yet another satellite, to be rebroadcast to as wide an area as desired. Small receive-only dishes at each of the remote sites receive the signal, decode it, and route it to a server outfitted with a plug-in DVB card. If BBI terminals are also located at the remote sites, then two-way exchanges of information can be implemented.

Such two-way satellite terminals can also provide a means of broadband access to the Internet for residential users. While the numbers in use for this purpose are relatively modest at present, it is expected that this will increase greatly in the future as the cost of two-way terminals is reduced. Figure 5.3, for example, shows the basic principle involved with using a BBI terminal for Internet access.

A BBI data interchange involves a so-called "double hop," which gives rise to the time delay associated with two separate satellite transmissions. As was previously stated, while most services are insensitive to this, some, such as voiceover IP telephony or video conferencing, do not work well when the latency of a round-trip with double hops is introduced. Experiments have shown that the delay associated with a single hop goes almost unnoticed in such services, but a double-hop delay becomes quite annoying to users.

To avoid this problem, one can make use of a degree of digital processing onboard the satellite to perform some of the functions normally performed in the BBI hub. One type of limited OBP that can already accomplish this is the Skyplex® processing system embodied on several of the Eutelsat satellites. Skyplex is a device, developed by the ESA, that allows multiple uplinks from different users to be regeneratively combined onboard the satellite. A user with an MF-TDMA terminal uplinks data to the satellite. Onboard the satellite, the signals are multiplexed together with those received from other terminals and broadcast down in the TDM stream that can be received and decoded by any intended recipient in the beam. If that recipient also has an MF-TDMA uplink terminal, they can respond. In this way, only a single hop is involved in each direction and voiceover IP, for example, becomes easily usable.

While the Skyplex package can do this in a limited manner, a number of companies are developing systems, involving more extensive onboard processing,

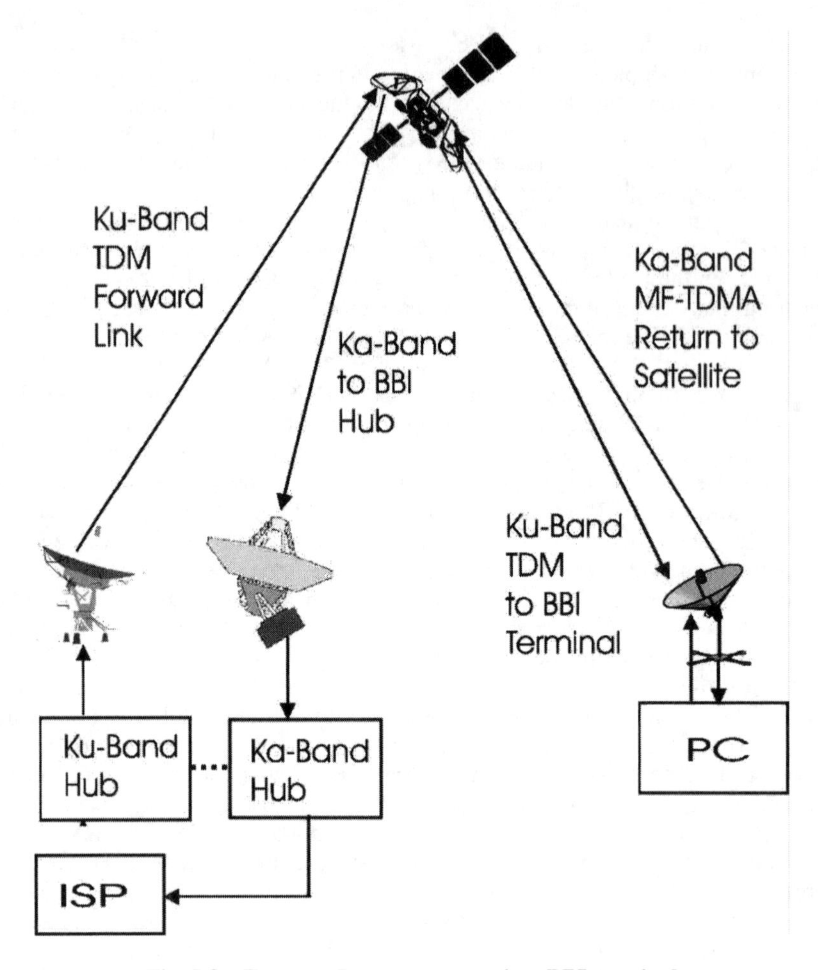

Fig. 5.3 Two-way Internet access via a BBI terminal.

that can both perform such functions and route signals from any receive beam to any downlink beam in a multibeam satellite system.

It is this latter type of mesh-connected service that caught the attention of the many in the satellite industry a few years ago. OBP satellite systems were proposed to provide broadband interconnectivity to millions of users. These included Teledesic, Astrolink, CyberStar, and Spaceway in the United States; West, SkyBridge, and Euroskyway in Europe; and WINDS and Gigabit Satellites in Japan. Even though carrying such one-to-one traffic is not the most efficient way of using satellite capacity to serve the largest user population, it is, for some users at least, the only feasible way for them to be able to obtain broadband access to the Internet. The terrestrial alternatives, broadband access via television cable modems or ADSL, are not available everywhere, and probably will not be for many years to come.

The bursting of the "Internet bubble" at the beginning of 2000 caused the disappearance, or significant delay, of most of these systems. Nevertheless, the market surveys that led the proponents and investors in these systems to initiate them did indicate that there was a potentially lucrative market for such systems (albeit perhaps, not when in competition with so many others) and this virtually untapped market will probably justify one or more such systems being developed in the not too distant future.

After broadcasters and operators broke out of the narrow confines imposed by the previous analog transmission standards, there suddenly appeared a virtual rainbow of possible applications and services that could take advantage of the new digital world. These started from enhancements and additional services related to television broadcasting, but soon expanded to areas completely outside the television domain. Operators at first spent time looking for what they called the "killer application," or "killer service," which would suddenly become in demand by practically everyone, and reap great rewards to the operator that first was able to offer it. It soon became apparent, however, that because the digital standards allowed such wide flexibility in what could be offered, there really was not such a single application. Many applications and services could be offered in parallel, and this is the path now being followed by many operators.

AstraNet, provided by the SES ASTRA satellite constellation, is being used in Germany to provide what is called the T-DSL Service to Deutsche Telekom T-Online subscribers. This is a service that operates in parallel to a two-way ISDN connection from a user to an ISP. It monitors the real-time demand for bandwidth on the terrestrial forward link. When this demand exceeds a certain threshold, a link from ISP to user going via satellite is automatically invoked. Data then goes to an SES receive terminal at the user's residence and is then combined with data arriving on the terrestrial link. This provides download speeds meeting or exceeding those available via ADSL connections, to users that are outside the areas where ADSL connections are possible. SES is also exploiting the AstraNet platform to support a similar service offered by a new Internet company in Italy, NetSystem.com.

The BBI system is being marketed for many different types of applications. At one end of the spectrum, the BBI system can operate together with the AstraNet system to provide multimedia content to the AstraNet hub, from where it can be aggregated and distributed to users via AstraNet. BBI can also be used to provide wider bandwidth services to business users than their present VSAT systems are capable of. It can also be used to interconnect cable head ends to grant their users access to the Internet via their cable modems. The possibilities are almost endless.

B. Multimedia Reception Enhancements

The DVB standards have also made possible the establishment of standards and specifications for what is called the multimedia home platform (MHP). Up until now, service providers developed their own applications for their own platforms, using application program interfaces (API) that were proprietary to their system. In general, this meant that their systems were not interoperable with those of other service providers. This had the particular downside that manufacturers had to build different terminal designs to different proprietary standards for each operator, and could therefore never fully achieve the economies of scale that a truly mass market could bring. The establishment of the DVB standards has changed this, at least in part. Agreement has now been reached in Europe on an MHP standard so that all service providers can write applications, through a common API, for use with the MHP platform. The MHP standard, which was approved in 2001, covers a wide range of potential terminal functionalities, and includes all needed protocols, languages, and interfaces to support both broadcasting and Internet access services. It is very possible that it may revolutionize the fields of both satellite and terrestrial multimedia service provision, not only in Europe but also throughout the world.

C. Video on Demand

For a time, there was considerable "hype" that broadband satellite systems would be able to provide video on demand (VOD) services to millions of viewers at prices per film comparable to or less than the prices paid to rent videos or DVDs at a local video store. However, some simple sums done as a sanity check to this proposition can reassure proprietors of the video stores that they can breathe a bit easier about their futures.

With MPEG-2 compression, a VHS-tape-quality film can be transmitted with a bit rate of some 2 Mbps. Considering that a typical film averages 2 h, such a film could be transmitted in 2 (Mbps) $\times$ 2 (h) $\times$ 3600 (s/h)/8 (bits/byte) = 1.8 GB. If this film were sent over a typical 27 MHz wide satellite television transponder, having an effective spectral efficiency of 1 bps/Hz, it would take approximately 9 min. Approximately 6.66 such films could be sent in 1 h, and if the transponder in question was fully occupied for an entire year doing nothing but downloading such films continuously, approximately 58,500 films could be sent. Because a typical television transponder leases for $1.5 million per year or more, with higher than prorated prices for short-term use, the transmission cost alone for each film would wind up being more than $26. Obviously, not many people would want to order a film via satellite when they could rent it for approximately one-tenth the price at a local video store. Such point-to-point film delivery would not be competitive at all.

In the past few years, the cost of computer hard disk storage, on a price per gigabyte basis, has been decreasing rapidly. At the beginning of 2000, retail prices ranged between $10 and $30 per gigabyte of storage, for hard disks ranging up to approximately 20 GB of total storage capacity. By mid 2002, prices had fallen to between about $1 to $3 per gigabyte, for disks ranging from 40 to 80 GB of capacity. While there are physical limits that will keep this trend from continuing indefinitely, one can be rather confident that disks with storage capacities exceeding 180 GB will be available within a few years, at prices less than $100.

The ability to store vast amounts of digital data in the home cheaply, in a PC or set-top box, will make it possible to receive and store perhaps 100 or more films at any one time. These could be broadcast by satellite to multiple subscribers in nonpeak hours (say between midnight and 0600) over a period of several days, and thereafter be available for viewing on demand from the subscribers' own disk store. In this way, by multicasting films to thousands of subscribers at the same time, the effective transmission cost per film can be reduced to only a few cents.

Of course, the transmission costs are only one set of costs in the film rental chain. The chain includes the film studios that own the films, the video service provider, the satellite operator, and perhaps several other retail or wholesale steps before a subscriber receives a film at home, and each has to receive its share of the overall costs of such a VOD service. To ensure this, and to avoid films being received by pirates that are not subscribers to the service, the films sent will of course be encrypted. Each time a subscriber wants to actually view a stored film, he will have to request a unique code to decrypt that specific film. The VOD supplier will then debit the subscriber's account for an agreed amount that will be sufficient to provide the profits needed by each element in the distribution chain.

A number of different schemes can be considered for updating the films on a subscriber's hard disk. Perhaps the simplest system would be to update the whole set at certain periods, perhaps monthly. Alternatively, the subscribers system could have software allowing the purge of any films from the hard disk that are not of interest or that have already been viewed. The same software would then fill up the empty space thus made available with new films that could be in the bouquet of films multicast to subscribers every night.

Whatever the charging and updating scheme chosen, however, it appears likely that such VOD services via satellite will become available in the not too distant future, and that they will be priced competitively with the local video stores.

D. Movie Delivery to Theaters

It is estimated that there are some 25,000 commercial movie theater screens in operation in the United States, approximately 20,000 in Western Europe, and perhaps as many as 5–10 times these amounts in the rest of the world. Up until now, the projectors for almost all of these screens had to be provided with several large canisters of celluloid film for each movie to be shown. The cost of making multiple celluloid copies of a movie and transporting one to each theater where it is to be shown is estimated to be approximately $1,000 per copy. Even though not all theaters get the same films at the same time, so that the celluloid film copies can be used for a number of theaters in a row, the overall infrastructure cost to copy and distribute films is impressively large. Steps are being taken, however, to reduce these costs significantly.

The technology has now become available to store movies digitally on magnetic tapes, which are far smaller and cheaper than the celluloid film equivalent. Projectors are also now available that can accept these tapes and project bright, high-resolution movies onto large theater screens. A new technology projector, such as this, is admittedly far more expensive today than a standard projector, but it is expected that the cost differences will lessen with time as more digital projectors are produced. Furthermore, because the digital magnetic tape needed to store a movie costs an order of magnitude less than celluloid film, for the same movie length and quality, and is reusable multiple times, the cost advantage of the digital technology will, as more and more movies are shown, gradually outweigh the initially higher price of the digital projector.

It thus seems likely that there will be a gradual shift from conventional projectors to digital projectors. It remains to be seen, however, whether the distribution system in existence today for celluloid film reels (the mail, or commercial courier services such as DHL or Federal Express) will also give way to the use of electronic means.

A movie with sufficient resolution to be viewed in a large widescreen theater requires far more storage capacity than the VHS video discussed in Sec. IV.C. Even compressed with multimedia MPEG-4 algorithms, which generally offer higher compression ratios than MPEG-2, a high-quality, large-screen film can occupy 25 Gigabytes or more. With the same assumptions for the transponder considered for the VOD case, this would take a bit more than 2 h to send via satellite. The cost for this transmission would exceed $350, which would not make distribution by satellite on a point-to-point basis any better for films than it was for video. However, the same solution can exist, as was the case for VOD. If the film were sent to only ten theaters simultaneously, rather than to one, it would cost no more to send, and the cost per theater would then reduce to approximately the same as it would cost to send a digital tape by courier. In this case, satellite distribution could begin to have a viable business case. Furthermore, for new releases, many hundreds or even thousands of theaters could benefit simultaneously from satellite distribution, making the satellite even more competitive.

For businesses such as theaters, however, it could be that competitive alternatives exist in many cases. While it will probably be a long time before all theaters have direct fiber access, most could have an ADSL telephone connection, allowing up to (at least theoretically) 8 Mbps of downloads. With ADSL, because of its generally more benign noise and interference environment compared to satellite transmissions, higher-order modulation schemes could be used to perhaps double (or possibly even triple) the spectral efficiency compared to that of the satellite. Thus, downloading a film on an ADSL link would take perhaps the same time, or only a bit longer than via satellite.

Would ADSL then win out in every case? This is not clear. An 8 Mbps ADSL link is not available everywhere (most telephone companies offer a maximum of 1.5 Mbps) and when available does not cost the same as the typical links often available (in 2002) to residences. If the higher-capacity ADSL services are available at all, one can expect to have to pay 10 to 20 times the typical monthly residential rate, and this is independent of how much time the line is actually used for downloading movies. Many theater owners could think this an expensive proposition.

Will satellite delivery win out in every case? Probably not. Many small, single-screen theaters may be quite content with receiving 8 to 16 films each month by courier and see no good reason to change.

E. Interactive Television (iTV)

Watching television has always been a rather passive and sedentary activity, if one excludes people getting up during commercials to fetch snacks or drinks, and those people that watch and take part in aerobics and physical exercise programs. It will probably always remain sedentary for the most part, but there are trends that indicate that it could become much less passive in the future.

A number of cable and DTH programs are now being broadcast that solicit or allow a degree of participation from the viewers. These include, for example, pay per view (PPV) films, where the viewer has to place a telephone call and agree to pay to see one of a number of films that may be scheduled for broadcast shortly thereafter. Tele-shopping is another example, where programs solicit viewers to call in and order one of the products being discussed or demonstrated in a broadcast.

Interactivity is made easier for viewers where EPGs are downloaded to the set-top boxes or satellite decoders that can be accessed by "clicking" on the appropriate buttons on a remote control unit. In a similar fashion, some DTH systems provide the capability for viewers to merely click on a particular button or buttons on a remote control unit to become enabled to watch a PPV movie. This is the case in the United Kingdom, for example, for the SKY Digital programs. The set-top-box or satellite decoder for SKY broadcasts has a memory that will record the fact that a viewer has pressed the appropriate button(s) to view a particular movie. It has, as well, a direct connection to the PSTN to allow it, rather than the viewer, to place a phone call and automatically give the SKY back office computer the information to enable the charge for watching that film to be included in the next monthly bill. The set-top box will even delay making this call until the early hours of the morning, to minimize the possible interference with residents wanting to use the phone at the same time for other purposes.

In both the EPG and PPV cases previously discussed, the real-time interaction is limited to that between a viewer and the set-top box itself. However, if the box had an "always-on" return connection to the outside world, many more types of interaction, interactive services, and applications can be envisaged. For example, voting in real time on the outcome of a televised talent or beauty contest or a figure-skating competition could be done by the population at large, rather than by a limited number of judges. Audience participation to quiz shows could expand to the entire viewing audience, not merely to the audience in the studio from where the show is taking place. Interactive games with players using their remotes to control computer-animated "Avatars" on the screen would be possible, and possibly these would become even more popular than their present PC game equivalents, where only two or a few people can participate. Of course, given the finite delay in sending signals via a geostationary satellite (one-half second for the round trip to and from the satellite), really fast action video type games would be difficult to play, but board games and role-playing games, to mention only two types, would work well. Finally, perhaps the biggest market possibility, in regions where it is legal, would be the application of interactive TV to gambling. This is already a big money earner on the PC-centric Internet, but could become much more lucrative if the broader television viewing audience were enabled to participate. Viewers could use their remote controls, for example, to place even last-minute bets on horse races, boxing matches, and football games, to name a few.

In addition to being able to provide additional types of services to viewers, an always-on connection can be very useful to broadcasters. In the first instance, it can allow broadcasters to "poll" terminals and have them automatically send back their serial and or account registration numbers. If an illegal terminal is detected, it can be excluded from the system. Second, set-top boxes can be designed to store information on what programs are being viewed, and to send this data back to the broadcaster at fixed intervals or when polled. This information can be used statistically to provide competitive ratings of the programs. Such user profiling also makes it possible to send targeted advertising to viewers, based on an analysis of what types of programs they watch most.

In making use of the information provided to them by always-on connections, broadcasters will have to be careful not to violate privacy and data security laws that have been passed in a number of areas of the world. If the profiling is done at set-top box level, however, and only used together with built-in software to filter out which received transmissions are stored on the internal hard disk, there should be no problem because, in such a case, there is no transmission back to the service provider of the individual's preferences. Likewise, if preference information concerning what programs are being watched is sent back to a broadcaster, but without an identification of the individual having such preferences, personal privacy laws will probably not be violated. Nevertheless, broadcasters will have to be careful to err on the side of conservatism when obtaining information from the set-top boxes.

An always-on return connection can be provided in at least three ways:

1) with a hard-wired solution such as a direct connection to the PSTN,
2) with a wireless connection through a cellular phone network, or
3) with an RF return link by satellite.

The direct connection to the PSTN is conceptually simple because most people who watch television also have a telephone, but it has several drawbacks that tend to inhibit its popularity. First, it requires connecting wiring from the set-top box to the nearest phone jack, which in some cases may be on the other side of a living room or even in another room in the house. Second, when the phone line is used for iTV, it is not available for other calls, either in or out. Of course, a homeowner could have another PSTN line installed, or opt for an ISDN connection if available because it typically has two lines. In such cases, the iTV would only block one of the two lines, but both solutions cost more money, both initially and on a monthly basis. Finally, in most countries, subscribers are charged for each call, even if it is a local call, so these solutions could result in significantly larger (and difficult to predict) phone bills for a viewer.

Using a wireless return connection with, for example, a GSM chip built into the set-top box could avoid some of the difficulties considered for the PSTN, but a cellular phone call typically costs significantly more than that from a normal phone. Also, with most cellular systems, the connection is not really actually always on, and a possibly objectionable delay would be experienced every time a viewer wanted to interact.

Development of a low cost, very low bit rate return capability for DTH terminals is a goal currently being pursued by a number of companies and research facilities. This would allow a user to interact with a television program directly, via the satellite over which the program is being broadcast. The idea of a low-cost transmitter to send signals to a satellite nearly 40,000 km away may, however, seem like an oxymoron. Conventional VSAT transmitters can cost many hundreds or even thousands of dollars. These are, however, designed to be able to send relatively high data rate signals to a satellite, which in turn means that they have to have a relatively high RF output power because there is a direct relationship between data rate and output power required. Likewise, there is a direct relation between output power and transmitter cost.

Fortunately, the amount of information that must be sent with each click of a remote's button in an iTV system is very small. Only a few tens of bits need to be sent, for example, to provide a viewer's account identification number and to identify the program that is being watched and the item that the viewer wishes to purchase. This can be sent in a very short burst at 1–10 Kbps, for example, rather than at the hundreds (or even thousands) of Kbps often associated with VSATs. Such a low bit rate burst can be sent via satellite using transmitters that output only a few tens of milliwatts, instead of the tens of watts that may be associated with high-data-rate transmitters. At such low power output levels, transmitter costs can be as little as a few tens of dollars, rather than many hundreds of dollars, so the oxymoron previously mentioned may be more imagined than real. Furthermore, because any user will only be sending a few tens of bytes at any one time, many thousands of users can share the bandwidth of a single transponder, making the cost of such a return channel easily affordable to the average consumer.

F. Broadcasting to Mobile Terminals

The services previously described have all been concerned with broadcasts to (and/or returns from) fixed terminals. Fixed terminals working in conjunction with GEO satellites have a great advantage, in that their antennas can be aimed permanently toward a single and fixed location. Furthermore, fixed terminal antennas can be relatively large (60 cm in diameter or more), and thus have a relatively high gain in the desired direction that enables them to reject interfering signals from other satellites.

There is, however, a lot of development work that has been undertaken, and is still going on, to allow mobile terminals to receive satellite broadcasts as well. Broadcasting, and even interactivity via two-way connections are being considered for cars, buses, trucks, trains, and planes. Even future generations of handheld personal phone terminals could become a market for satellite broadcasting.

This is not really anything particularly new, of course. Inmarsat, for example, has been providing communications to and from mobile terminals for years, and making a good business of it. Communications, however, is not the same thing as broadcasting. Inmarsat has been operating in the L-band spectrum allocated by the ITU for MSS, but the bandwidth available is very limited. Inmarsat only has access to some 24 MHz of bandwidth for its satellite-to-ship links, which is enough for approximately 4800 simultaneous telephone calls in the region covered by one of their satellites (close to 1/3 of the globe). This same bandwidth could also be used on a shared basis with a broadcasting service, but even the total bandwidth is not enough to do any extensive broadcasting. Furthermore, up until now, Inmarsat has obtained the majority of its revenues from communications with terminals mounted on ships and boats, and to a much lesser extent, on airplanes. While Inmarsat does provide voice, fax, and low-rate data services to small briefcase-sized terminals that can be carried and used by people on land, these have normally only found use in areas of the globe not served, or at least inadequately served, by terrestrial communications means. In areas well served by cellular or fixed line services, the relatively high prices that Inmarsat must charge to make money with such a little bandwidth make it difficult to be competitive.

Inmarsat is, however, developing a fourth generation series of satellites, which, beginning in 2004, will use hundreds of narrow spot beams to allow multiple reuse of its narrow frequency allocation. These satellites will enable it to provide higher data rate services than at present, at prices well below those being charged for capacity on the Inmarsat-3 series and previous generations of its satellites. This will allow much more effective use of bandwidth for communications. Multiple small spot beams, however, are not an efficient way to broadcast signals over wide areas, because the frequency reuse capability provided by the multiple beams is eroded when the same signals must be sent on many beams. Thus, perhaps ironically, even though the total capacity of the fourth-generation Inmarsat satellites will be many-fold times that of its predecessors, the amount of capacity that can economically be devoted to broadcasting may not be significantly increased. An exception to this could be narrow-casting, or broadcasting to a number of terminals all located in the same spot beam. Here, the increased equivalent isotropic radiated power (EIRP) that the spot beams will produce will make it possible for broadcasts to be received by terminals with relatively small antennas, while still achieving a respectable data rate.

G. Sound Broadcasting

One area where satellites may be successful in operating in the L-band and S-band, where only relatively narrow frequency allocations are available for such a service, is in direct broadcasting, to relatively large areas, of radio programs to cars and trucks. Audio transmissions require far less bandwidth than those for television. Terrestrial FM radio stations can provide high-quality music broadcasts, but they are limited in range, and many people feel frustrated at having to retune their car radios to another station every 20–30 min when driving on the highway. AM, and in some places long wave (LW) radio stations do broadcast over larger areas, but their quality is poorer and subject to more interference than FM stations.

Broadcasting to terminals mounted on cars and trucks presents difficulties that do not occur when using terminals fixed at a specific location. The first of these is the fact that it is impractical to use large diameter parabolic dishes mounted on such vehicles. A 24-in. or 60-cm-diam dish mounted on the top of a fast sports car (or even on a beat-up old VW) does not exactly add to its streamlined shape, to say nothing about what impact it might have on the aerodynamics and average fuel consumption at high speeds. Much smaller antennas must be considered for such an application, which necessarily cuts down on antenna gain, thereby either requiring higher-power broadcast satellites or putting limits on the bandwidth of the signals that can be received.

A second problem posed by terminals mounted on moving vehicles is that they change direction constantly. While a satellite dish mounted at a home site for TV reception can be accurately aimed toward the satellite from which it is to receive, and then be bolted in that orientation, a vehicle mounted antenna either has to be nearly omni-directional, or it must have a tracking capability to be able to maintain its boresight direction pointed toward the satellite it is using. Nearly omnidirectional antennas have even less gain than small parabolic dishes, putting even greater restrictions on the bandwidth of the signals that can be received. An antenna to be mounted on the top of a car, which can rotate in azimuth and point in elevation, is both complicated and expensive.

A third problem, illustrated in Fig. 5.4, is that, unlike ships and airplanes, cars often pass behind or underneath obstructions such as buildings, trees, tunnels, bridges, and even mountains. While fixed-terminal antennas can usually be located to avoid such obstacles, for mobile terminals these can cause signal fading or outages because of the blockage of the signals from a satellite, or introduce interfering reflections. The latter gives rise to multipath interference, where the receiver receives both the desired signal and one or more time-delayed (because they must travel further) reflections.

There are ways that have been developed to reduce the effects of fading and blockages. For example, the same signal can be broadcast twice from a satellite, with the second transmission delayed several seconds in time after the first. The first of these can then be stored at the receiver and only played out some seconds after it is received. If that signal is lost momentarily, when the car

Fig. 5.4 Satellite signal obstructions.

drives under a covered bridge, for example, the second signal, received when the car exits from under the bridge, can be used to "fill in" for any lost data in the first signal. Of course, this only works if the duration of the interruption in the first signal is less than the time delay between the transmission of the two signals from the satellite.

Another way to reduce some mobile outages is to use more than one satellite for transmissions. If two GEO satellites are used that are relatively widely separated in longitude, the chances of the signals both being blocked simultaneously by a tree or a building are reduced. This approach is used by the XM satellite system deployed in 2001 in the United States.[7] Of course, this helps for relatively small obstacles, but cannot cope, for example, with tunnels. If, however, the trick of staggering the transmission in time through the two satellites is used, as previously described, even short tunnels and bridges can be accommodated (provided that traffic cooperates).

For mobile terminals in relatively high northern (or southern) latitudes, the effective elevation angle from the terminal to a geostationary satellite may be rather small. In such cases, even rather low buildings or trees can cause signal outages. The elevation angle of a terminal in the United States aimed toward a GEO satellite can vary from a high of approximately 55 deg in southern Texas to less than 35 deg along the Canadian border.[7]

One way to reduce such problems is to use a constellation of a number of satellites, arranged in orbits so that at any one time, one or more of the satellites will be visible at relatively high elevation angles. This is the system concept chosen by the Sirius satellite system deployed in the U.S. in 2002.[8]

Reference 7 discusses the amount of signal power margin required onboard a satellite to achieve a particular level of availability for various elevation angles. It shows, for example, that for L-band downlink transmission in the United States, with an elevation angle of 35 deg, as much as 23 dB of margin is required to overcome the blockage and fading because of foliage in rural areas to achieve an availability of 99%.

Only 12 dB of margin is required, however, to achieve an availability of 90%. Therefore, if the same signals are transmitted from two satellites that are far enough apart in azimuth to be able to consider that their probabilities of blockage are statistically uncorrelated, only 12 dB of margin would be required on each satellite to achieve an overall availability of 99%.

At an elevation angle of 55 deg, which is typical of that in the U.S. from the Sirius satellites, only a 12-dB margin is required to achieve an availability of 99%. Thus, one satellite (at a time) can provide essentially the same availability in the Sirius system that requires two satellites in the XM system. There is, as they say, no such thing as a free lunch, however, for in the case of Sirius, three satellites are required in non-GEO orbits to ensure that there is always one providing high-elevation angles.

Finally, whatever the means chosen, there will be times and locations, particularly in urban areas, where signal loss from the satellite(s) is unavoidable. Both the XM and Sirius systems therefore use terrestrial repeaters in the cities in which they operate to provide additional continuity of service.

While the systems discussed are targeting broadcasts to cars as their primary market focus, they could very well wind up with a sizable number of fixed radio subscribers as well, particularly from rural areas that lack the FM radio infrastructure that predominates in the cities. One company, WorldSpace, is already attacking this market in Africa and Asia, but with relatively little financial success thus far. However, when satellite technology can serve both potential markets, fixed and mobile, from the same satellite system, the synergistic effects may result in a surprisingly good business case.

H. Aeronautical Broadcasting

At any given instant, there are many thousands of commercial airliners in flight around the globe. Each of these contains a "captive audience" of travelers that airlines try to keep happy through the provision of in-flight entertainment. Most long-haul airplanes show one or more motion pictures during the flight. Even short-haul flights often show short news films or documentaries. Satellites now make it possible to broadcast to airplanes in flight. Thus, travelers will be able to watch television news as it happens, as well as watch sporting events in real time. While this is only possible on a limited number of flights at present, it can be expected to expand greatly in the future, not only to many more flights, but with a much wider selection of onboard television viewing possible.

With the advent of the Internet, another type of in-flight entertainment has become of interest. Many airlines are considering providing the means for their passengers to access the Internet, either using their own portable computers, or through a system using a screen mounted in the back of the seat and a small keyboard. Systems are being investigated, and some are already entering service, to allow Web surfing, at least in a limited fashion, and access to personal e-mail files. Both one-way and two-way systems are being considered.

All of these new aeromobile applications, however, rely on satellites to provide the needed communications capacity. Initially, satellites using L-band, such as the Inmarsat fleet, will be used for such systems. For the reasons previously discussed, however, this is only an interim step. If the market for two-way aeronautical services grows at the rate being predicted by some, the bandwidth available at L-band could soon prove to be insufficient, even when combined with the multiple frequency reuse capabilities of satellites such as Inmarsat-4. Eventually, higher-frequency bands will have to be used to meet the expected demand.

I. Convergence of Navigation, Communications, and Broadcasting

The GPS and GLONAS satellite systems of today make it possible to find one's location practically anywhere on Earth. Furthermore, navigation terminals have become inexpensive enough that practically anyone in the developed world can now afford to buy them. More and more cars are coming equipped with built-in satellite navigation systems. Practically all private airplane and boat owners now have installed satellite navigators. It is big business, but it could become a lot bigger.

Knowing where you are is a comforting feeling. After all, no one likes to become lost. That feeling of knowing where you are, however, benefits only one person—the person with the navigation terminal. If you (or the navigation terminal itself) can communicate your location to your employer, colleagues, spouse, children, or friends, then many more people can reap the benefits of satellite navigation. Moreover, communication of one's location opens the door to a variety of other types of services that can improve efficiency, safety, and even enjoyment. Emergency rescue, hazardous cargo tracking, stolen vehicle recovery, targeted road condition reporting to motorists, and targeted advertising all become feasible or easier to implement. In certain locations, such services are already in operation using terrestrial communications links for position reporting, but the use of satellite for this purpose can make them more ubiquitous, as well as more affordable.

If we add the third dimension of being able to "narrowcast" information to vehicles that is related to their location at that time, an even wider pallet of new types of services can be envisaged. Local weather and road and traffic conditions (such as road work or traffic jams) can be valuable to motorists in enabling them to complete their journeys more quickly and safely. If such local information is narrowcast to the in-car navigation systems that many cars now come equipped with, intelligent rerouting can even remove the burden from the driver of a car of having to determine how best to avoid such obstacles.

Narrowcasting based on vehicle location also allows targeted advertising, informing motorists, for example, of filling stations, hotels, or restaurants ahead. The two-way communications capability that satellites can provide can also allow motorists to respond to such advertisements by booking rooms or making reservations in restaurants ahead.

V. Summary and Conclusions

Satellite broadcasting systems as we know them today are in a state of flux. They are evolving in the face of the changing business and competitive environment, and improving through the incorporation of the new, higher performance technology being developed for both the space and ground segments. Digital broadcasting and satellite access standards are also in a state of flux, but here there is at least a possibility of convergence in the future, with one world standard finally arising. This will allow the production economies that mass markets bring to reduce the price of, in particular, user terminals, which should open the door to significant increases in subscribers.

Additional services will become available in the near future that will increase the attractiveness of satellite content delivery and access in some marketplaces, but there is the ever-present danger that the outgrowth and improvement of terrestrial systems will begin to marginalize satellites in other markets. A number of these new services have been addressed in this chapter, but this is only a sampling that shows the "tip of the iceberg." Many other services are possible, such as distance learning, various financial services, online voting in local and national elections, films with interactive plots, profiling of set-top box usage to provide targeted advertising and "push" information downloading, and "micro-broadcasting" where everyone can become a broadcaster. The possibilities are enormous.

Finally, one recurrent theme that has been lying just under the surface throughout this chapter is the theme of convergence. There are chapters in this book devoted to fixed, mobile, and broadcasting services, but the distinction between these is rapidly beginning to disappear. As was said earlier, "bits are bits," and after everything in the telecommunication satellite world is transmitted digitally, there is really no distinction between the three service types, other than that due to the environment in which they operate. Even the environments, because of advances in technology, are growing closer together.

References

[1]Long, M. E., *The Digital Satellite TV Handbook*, Newnes, Oxford, 1999.
[2]Baylin, F., *Digital Satellite TV*, Chapter 4, Baylin Publications, Boulder, CO, 1997.
[3]SES GLOBAL Annual Report, Betzdorf, Luxembourg, 2001.
[4]*Interavia*, No. 656, Sept. 2001.
[5]Futron Corporation Executive Summary, "Satellites Steer Into the Future: Course is Strong, if Not Steady," 12 Aug. 2002.
[6]Cáceres, M., "Orbiting Satellites: Bean-Counter's Heaven," *Aerospace America*, Aug. 2001.
[7]Michalski, R., and Nguyen, D., "A Method for Jointly Optimizing Two Antennas in a Diversity Satellite System," *Proceedings of AIAA 20th International Communication Satellite Systems Conference*, AIAA, Reston, VA, May 2002; also AIAA Paper 2002-1996.
[8]Briskman, R. D., and Sharma, S. P., "DARS Satellite Constellation Performance," *Proceedings of AIAA 20th International Communication Satellite Systems Conference*, AIAA, Reston, VA, May 2002; also AIAA Paper 2002-1967.

Questions for Discussion

1) Cable television is often transmitted with high-order modulations such as 64 QAM or more. Broadcasting satellites of today generally use QPSK, and hence have significantly less spectral efficiency than cable. **Comment on why satellite television broadcasting, in spite of this advantage of cable, is still growing faster than cable television in most regions. Make a comparison of the parameters influencing the economics of the two types of television media, and draw your conclusions.**

2) **Use library and Internet information resources to compare the image quality of television transmissions via NTSC, PAL, SECOM, ATSC, MUSE, and DVB-S. Compare these in turn with the image quality obtainable from VHS tape and DVB films. What conclusions can you draw from these comparisons?**

3) Tele-education, or the teaching of classes where the teachers are remote from the classroom and telecommunications links are used to give the teachers a "virtual presence" there, has often been quoted as a "natural" application for satellite broadcasting. Nevertheless, this type of application is still only available to a very limited scale, and only in certain areas. **Why do you suppose this is so, and what could be done to correct any problems that limit its use?**

4) The continual expansion of hard disk capacity, with constantly reducing costs per megabyte of storage, is quoted as a prime reason why Personal

Video Recorders (PVR) will become a rapidly expanding market in the future, which will primarily be oriented toward PVRs being able to download and record multiple films, and make them available in pay per view systems. There are, however, objections to such a service coming from the film studios, the providers of the films, who fear that it will allow piracy of their films. **How might such piracy be minimized or eliminated? If the film studios concerns prevail, what other types of services can you envisage that could also expand the market demand for PVRs?**

Chapter 6

Technical Profile of Next Generation Satellite Technologies

Hiromitsu Wakana* and Takashi Iida†
Communications Research Laboratory, Tokyo, Japan

I. Introduction

THE advancement of technology becomes faster and faster, especially in the information technology (IT) era. However, the rather constant pattern of development of communications satellite technology and systems during the last 40 years has been defined by several key trends:[1]

1) the development of higher-gain and larger-aperture antennas,
2) the development of higher-performance, longer-lived, and larger power systems,
3) the ability to expand available frequencies for satellite communications either through spatial or other reuse techniques and/or the allocation of new and typically higher frequencies, and
4) the development of new orbital configurations or satellite systems with substantially longer life.

In the near-term future, improved modems, new materials, and improved solar cells could help to advance the field. However, the most important paths forward in the satellite field will be driven by either the ability to exploit new extremely high frequency (EHF) and millimeter wave frequencies or by new types of satellite designs that allow major "sea changes" or radical breakthroughs in satellite cost efficiency and also enable the implementation of user microterminals that are low in cost, highly compact, and user friendly. Clearly, new technology that allows the use of higher frequencies in the Ka-, Q/V- and W-bands would allow satellite systems of the future to support new broadband services and, potentially, to lower the cost of user terminals and overall satellite systems.

This chapter discusses, in the first instance, technology research and development (R&D) for the future communication satellites from the viewpoint

*Director, Yokosuka Radio Communications Research Center.
†President.

of an overall high-technology R&D program, particularly if it is one conducted by the government. It then enumerates the items of technology R&D needed to realize future communications satellites. Finally, a more detailed description is given for specific highlighted technologies.

II. Method of Technology R&D for Future Communications Satellites[2]

R&D for future communications satellites typically is part of an overall high-technology R&D program. Therefore, communication satellite R&D should be synergistic to and/or collaborative with other science and technology R&D. Thus, the preferable method of progressing the communications satellite R&D should contain four goals.

First, the communications satellite R&D program should be attractive not only for other space programs, but also for other science and high-technology R&D programs. Such other space programs would include those devoted, for example, to IT, nano technology, and bio-related technology.[2] We should recognize that space development programs can be formulated to promote such a science and technology R&D.

The second goal is to develop cost-competitive systems for the commercial world. The commercial world contains satellite communication facilities and terrestrial facilities. Communication satellites should be both complementary to and competitive with the terrestrial systems.

The third goal is to foster advancement in many fields by spinoffs in terms of technology and science, social science, and venture businesses. Space systems, including communications satellites, consists of total technologies. That is, they include many different technologies with wide applicability to other fields. This means there are many seeds to create venture businesses.

Finally, joint development with private enterprise should be promoted much more. Further, the promotion of a joint study with various countries in the world is important and should also be promoted.

III. Items of Technology R&D

A major goal being pursued by many R&D programs all over the world is to develop cost-efficient broadband satellite systems to provide an “Internet in the sky” capability. Technology developments will be needed in a wide range of fields to realize such an Internet satellite. Examples of the technology that should be developed are explained in the following sections.[2,3]

A. Pursuit of New Satellite Communications System

Chapter 7 will describe the concept of an Internet satellite only from the required communication capacity point of view (see Table 7.1). To develop the total system, many new concepts or ideas can be developed for future systems. For example, new systems integrating satellite and high-altitude platform systems (HAPS) are expected to become important for future very high data rate systems using optical communications.

B. Development of New Frequency Bands and Optical Communications

There are two types of communications links—one connecting satellites with Earth terminals (the space–Earth link), and the second connecting two satellites (the space–space link). The high-capacity link can be realized by the optical link for the space–space link. High-capacity space–space links can be realized using optical communications technology. However, optical communications links are not easily realized for the space–Earth link, because of the absorbing effect through the atmosphere and attenuation by clouds. RF links must therefore be used for this link. However, to obtain usable bandwidths on this link that are comparable with those obtainable in space–space optical links, frequencies higher than Ka-band must be used. The millimeter wave band must be developed. The V-band (50/40 GHz band) and W-band (80/70 GHz band) are candidates for this development.

Optical communication is, however, the ultimate system to achieve very high capacity. In the future, 5 Gbps, 10 Gbps, and even hundreds of Gbps will be realized. To use optical methods for a space–Earth link, space diversity or Earth stations in the desert area will be useful. Moreover, HAPS will be also useful as a relay station for the space–Earth link. This might be a solution for the very high data rates expected for third-generation Internet satellites.

C. Advanced Antennas

Antennas are always a key technology for satellite communications systems as long as radio waves are used. The proposed systems to be discussed in Chapter 7 require extremely high data rates so that very high gain antennas (large-scale antennas) and multibeam antennas are necessary. In particular, the super multibeam antenna is a key technology for future satellite communications.[4] The following three types of advanced antenna technology can be enumerated:

1) super-large-scale deployment antenna,
2) realization of a super-large-sized antenna by a formation flight of a large number of satellites, and
3) super-multibeam antennas created by a new beam formation method.

D. Development of Ultra-High-Speed/Optical Transponders

To realize the high data rate system, a lightweight and ultra-high-speed transponder is indispensable. In this case, a number of new devices and optical technology applications will be needed. We believe that nanotechnology related R&D will also be useful for this field.

E. Super Miniaturization of Satellite

Satellite weight is almost proportional to data rate, as further described in Chapter 7. However, launching very large satellites is very expensive and, beyond a certain point, is not feasible because of launcher payload mass constraints. Thus, lightweight bus technology will be key for the future extremely high data rate satellites. This technology will include lightweight solar power systems and lightweight bus subsystems, and nanotechnology can also be applied to good advantage in the future.

F. Earth Terminals

Advancements in Earth-station-related technology are also needed. The smaller the Earth station, the easier and less expensive the satellite communications will be. Development of bidirectional communications is needed, particularly those using nonsymmetric communication methods. Services other than communication should be included, such as positioning and satellite service functions. In the limit, for personal communications, super miniaturization of Earth station technology could even make terminals wearable.

IV. Highlights of Technology

In this section, more detailed descriptions are given for some of the technologies previously highlighted such as HAPS, more efficient orbit use, deployable antennas, active phased array antennas, ultrawide band (UWB) technology, and optical communications.

A. HAPS[5–8]

A stratospheric platform or HAPS, which is designed to stay at an altitude of approximately 20 km in stratosphere, has been conceived to provide communication, broadcasting, and environment monitoring services. Requirements for HAPS are 1) stay at an altitude of approximately 20 km and be operated as a repeater, 2) continuous operation for long periods, 3) capability of bearing large mission payload mass and providing enough power, and 4) environmental friendly equipment. Three categories of airships are candidates of HAPS—an unmanned airship with a propulsion system and a solar-cell-powered battery, an unmanned solar plane, and a manned aircraft as shown in Table 6.1.[6]

In 1998, a Japanese national HAPS project was started, led by the Ministry of Education, Culture, Sports, Science and Technology (MEXT) and the Ministry of Public Management, Home Affairs, Posts and Telecommunications (MPHPT). An airship-based HAPS was selected from the three candidates previously mentioned, primarily because of mission weight considerations and atmospheric environmental conditions. Table 6.2 shows specifications of the HAPS and the environmental condition of the stratosphere. Figure 6.1 provides a conceptual drawing of HAPS, which was produced in 1999 as part of the feasibility study by the National Aerospace Laboratory (NAL) of Japan. The body of HAPS is equipped with a number of helium-gas bags, a solar cell panel on the top, and three propulsion engines. Several hundred kilowatts of electric power are generated by the solar cells in the daytime, and are stored in an onboard battery for use at night. The lifetime will be more than 10 years, and during that period HAPS will return to the ground for maintenance several times.

A single HAPS can cover a service area with a radius of up to 100 km with elevation angles of more than 10 deg. This area is divided into smaller cells with radii of several kilometers by using multibeam antennas. HAPS platforms are

Table 6.1 Candidates for the stratospheric platform

	Large-scale airship, unmanned	Solar plane, unmanned	Jet plane, manned
Total length	200 m	70 m	30 m
All-in weight	30 tons	1 ton	2.5 tons
Propulsion energy source	Solar cell	Solar cell	Fossil fuel
Flight duration time	3 years	N/A	8 hours
Station keeping range	1 km	1.5 km	10 km
Mission equipment weight	1000 kg	100 kg	1000 kg
Mission allowable electric power	10 kW	1 kW	20 kW
System example	Japan, Korea, China, Sky Station	Helios, Pathfinder Plus (AeroVironment Inc., in the United States)	HALO (Angel Technology Corporation, in U.S.)

connected to each other via a point-to-point gigabit optical link to form a trunk subnet. Three service scenarios are considered—broadband wireless access in millimeter-wave for providing megabit-class, high-speed services to fixed terminals; compensative mobile communications services for IMT-2000; and broadcasting services such as digital TV and audio broadcasting.

1. Broadband Wireless Access Systems

Broadband wireless access will provide high-speed multimedia services to users with fixed terminals located in metropolitan, suburban, or rural areas. This will be an alternative solution to fixed satellite communications services. Even in the HAPS communications links, line-of-sight communication will be necessary,

Table 6.2 Specifications of HAPS and environmental conditions

Altitude	20 to 22 km
Maximum wind speed	30 m/s
Temperature	−56.5°C
Intensity of light	1.26 kW/m^2
HAPS length	245 m
HAPS weight	32,000 kg
HAPS propulsion method	Propellers driven by electrical motors

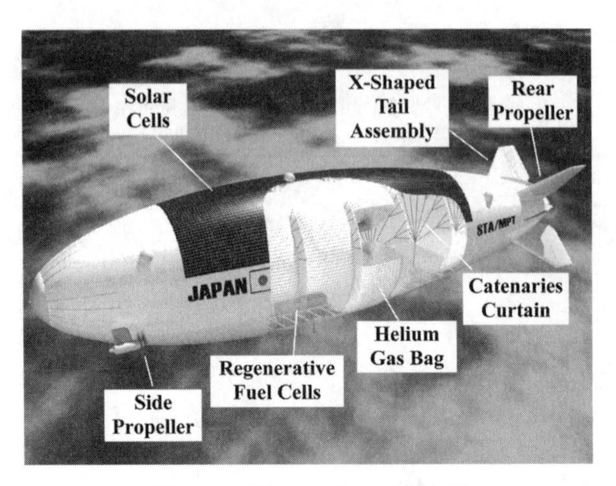

Fig. 6.1 Design view of HAPS.

and efficient modulation techniques should be used. Maximum data rates will be 100 Mbps in the downlink and 6 Mbps in the uplink. The use of the 48/47 GHz band and the 31/28 GHz band has been accepted already by the International Telecommunication Union (ITU) for fixed access.

2. Mobile Communication System Complementing IMT-2000

This is intended to allow HAPS to take over a number of the base stations of IMT-2000. The ITU has already approved part of the frequency bands assigned to the terrestrial IMT-2000 for use by a HAPS-based, IMT-2000 service. Two types of antennas, a switching planar antenna and a multihelical antenna, are being developed for HAPS. A user with an IMT-2000 terminal can access HAPS at 384 kbps.

3. Broadcasting Services

As for satellite, broadcasting services will be "killer" applications for HAPS. In Japan, terrestrial digital TV broadcasting services will be started in 2003, and will completely replace the present analog system by 2010. More than 10,000 terrestrial TV transmitting stations and repeaters are now operating in Japan. Satellite-based digital TV broadcasting services are already being provided all over Japan. HAPS will complement the operation of these TV broadcasting services. Moreover, digital audio broadcasting services, public information broadcasting services, disaster monitoring, and management services are also suitable for HAPS, although the frequencies of such HAPS-based broadcasting services have not been established yet by the ITU.

In the United States, several companies are preparing business using stratospheric platforms. Sky Station International, Inc., is preparing to offer a broadband fixed-access link of 2 to 10 Mbps in the 47/48 GHz band and an IMT-2000 link in the 2-GHz band using a 150-m class airship. NASA has developed an unmanned light aircraft powered by solar cells and has operated it at an altitude of 20 km. In July 2002, CRL and TAO of Japan carried out communication experiments using IMT-2000 terminals and HDTV broadcasting experiments in Hawaii, using the Pathfinder-Plus in cooperation with NASA and AeroVironment, Inc. Angel Technologies Corporation and Raytheon have a plan for providing communication services using a manned jet plane, named the HALO-Proteus. In 1998, an initial demonstration flight test was conducted at an altitude of 15 km.

In Europe, Advanced Technologies Group, Inc., in the United Kingdom is developing a stratospheric platform named StratSat, which is 200 m long and can embark mission payloads up to 2 tons. A combination of solar cells and a diesel engine are used as electric power sources.

Several organizations in Asian countries are interested in HAPS—Korea Aerospace Research Institute and Electronics and Telecommunications Research Institute in Korea, Shanghai Jiao Tong University, and Beijing Tsinghua University in China.

B. Orbit Use

As mentioned in Chapter 2, several types of satellite orbits are now being used. To describe a particular satellite orbit, six parameters are typically used. The origin is the center of the Earth, the z axis directs through the North Pole, and the Earth rotates on the z axis. The point in the orbit where the satellite is closest to the Earth is called the perigee, and the point farthest from the Earth is called the apogee. The eccentricity e indicates how elliptical the orbit ellipse is. The centers of the ellipse and the Earth do not coincide unless the eccentricity is zero. The length of a and b are the semimajor and semiminor axes of the orbit ellipse. At the ascending node, the satellite penetrates the equatorial plane from the negative z to the positive z. The right ascension of the ascending node is measured as an angle between the direction of the ascending node and the direction toward the first point of Aries. The inclination is the angle that the orbital plane makes with the equatorial plane. Another two parameters are the argument of perigee and the mean anomaly.

1. Low Earth Orbit and Medium Earth Orbit

The altitudes of low Earth orbit (LEO) and medium Earth orbit (MEO) are mostly in the range from 500 to 1500 km, and approximately 10,000 km, respectively, to avoid the van Allen belts. Most LEO/MEO systems such as Iridium, Globalstar, ICO, Odyssey, Teledesic, and SkyBridge use circular orbits, but some LEO systems such as Ellipso use elliptic orbits. Because the land and population of the Earth are asymmetrically distributed between the northern and southern hemispheres, the constellation of the Ellipso satellites is designed to offer efficient coverage. Europe, the United States, Canada, and Japan lie at 40°N, while southern New Zealand, Tasmania, Argentina, and Chile lie at 40°S, and all the areas further South of 40°S are relatively sparsely populated. The Ellipso system uses two planes of elliptical orbits inclined at 116.6° to prevent movement of the apogee on the orbital plane. The apogee and perigee are 7605 km and 633 km, respectively, with a 3-h orbital period. Values of the orbital parameters for several other LEO and MEO satellites are also given in Tables 4.4 and 4.6 in Chapter 4 as well.

Depending on the orbital radius, the orbital period and the satellite visibility time are changed. For example, LEO with an altitude of 1000 km has an orbital period of 1 h 45 min, and a satellite visibility time of approximately 12 min. Several tens of satellites would be needed to offer continuous global services. MEO with an altitude of about 20,000 km has an orbital period of 5 to 6 h and a dozen or so satellites would be needed for continuous global services.

2. *Highly Elliptical Orbit*

GEO, MEO, and LEO satellites typically have circular orbits with an eccentricity of nearly zero. Highly elliptical orbits (HEO) with eccentricity larger than zero can, however, be very useful for certain types of services. In GEO systems, the higher the latitude of an Earth station, the lower the elevation angle is to the satellite. In terms of satellite communications, particularly mobile satellite communications, blockage of satellite signals because of buildings, roadside trees, and utility poles, and multipath fading will occur more frequently as elevation angles become lower.

One solution to increase elevation angles in high-latitude countries is the use of an HEO. The former Soviet Union adopted this type of orbit, called "Molniya," that proved to be of practical use. The Molniya orbit has an eccentricity of approximately 0.5, an altitude of approximately 40,000 km at apogee, an orbital inclination of approximately 63°, and an orbital period of 12 h. Another type of HEO, the "Tundra" orbit, has a period of 24 h. Because the HEO is not a geostationary orbit, at least two or three HEO satellites will be needed to provide 24-h continuous services.

Another example of an HEO orbit system is the Archimedes[9] satellite constellation, proposed by the European Space Agency to provide good coverage of the densely populated northern hemisphere in Europe, and possibly in North American and Northern Asia as well. The Archimedes constellation consists of six satellites in different orbital planes, each having an 8-h period, with an operational orbit period of 4 h and an eccentricity of 0.63. The apogee and perigee are 26,786 km and 1000 km, respectively. During a 24-h period, one satellite can successively serve three Northern hemisphere zones spaced 120° apart in longitude.

3. *Geostationary Satellite and Quasi-Zenith Satellite (Figure-Eight Satellite)*[10]

A geosynchronous satellite has a constant angular velocity circular orbit with an eccentricity of zero. Because the orbital period of geosynchronous satellites should be one sidereal day of 23 h 56 min 4.091 s (86,164.091 s), the orbital radius is equal to 42,165.2 km, and its nominal altitude is 35,787.1 km. A geostationary satellite is a geosynchronous orbit with zero inclination, which means that the orbit lies in the Earth's equatorial plane. A satellite in the geostationary orbit appears to stay in a fixed position as seen from Earth stations. Because of simplicity in configuration, time-invariance of the satellite direction, extremely wide-beam footprint, and fixed propagation delay, GEO systems have been, and will continue to be widely used.

In land-mobile satellite communications in urban areas, blockage and shadowing because of roadside trees, utility poles, and buildings impair the communication links. One solution to this is the use of HEO, and another solution is a Figure-Eight satellite, which has an inclination of 45 deg but is geosynchronous because it has the same semimajor axis as the GEO. The locus of the subsatellite point traces a "figure eight" from the North to the South and vice versa, centered about a place at the equator. The Figure-Eight satellite can provide high-elevation angles even in medium-latitude countries like Japan. Because the orbit does not cross the van Allen belt, the degradation of the satellite-borne equipment because of the radiation is much smaller than HEO.

Figure 6.2 shows the locus of the subsatellite point of the orbit, which is optimized for the region at 35° North latitude in Japan, including Tokyo, Osaka, and Fukuoka. This optimized orbit has an inclination of 48.4° and an ascending node of 137.7°E. Three Figure-Eight satellites can provide 24-h continuous services, and a higher elevation angle than 70 deg in major cities along the Pacific Ocean in Japan and more than 60 deg nationwide in Japan. The three satellites are not orbiting in a single inclined orbit plane, but in three different orbit planes with

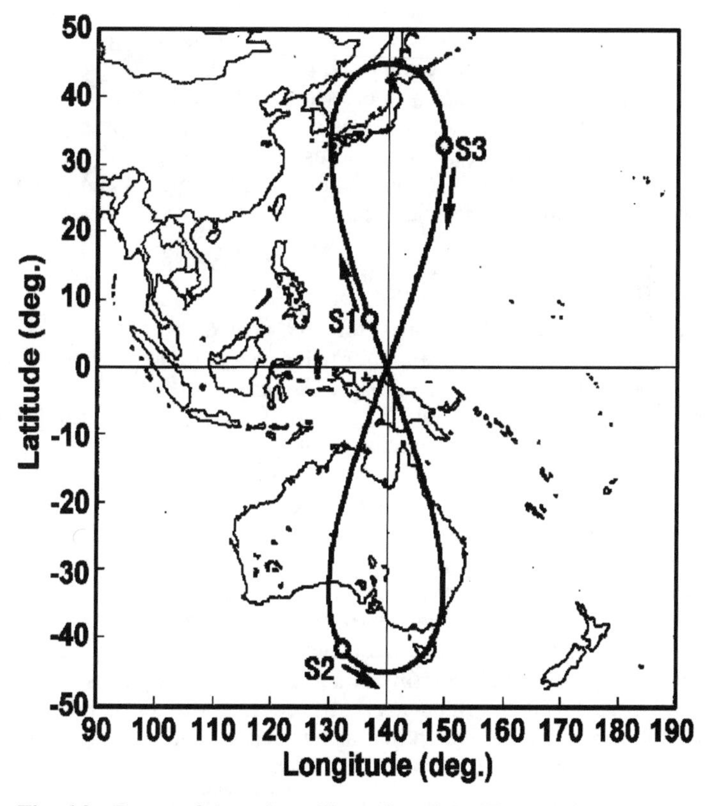

Fig. 6.2 Locus of the subsatellite point of the Figure-Eight satellite.

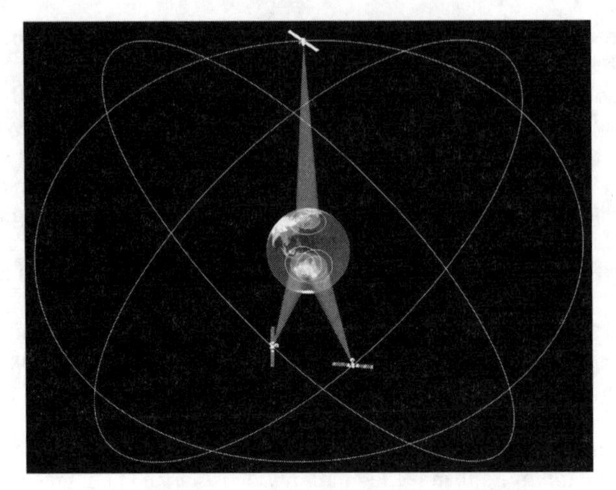

Fig. 6.3 Orbits of three Figure-Eight satellites.

the same inclination and ascending node intervals of 120 deg, as shown in Fig. 6.3. Each satellite moves on the locus with a separation of 8 h from the others. Continuous communication services can be provided by switching from one to another of the satellites every 8 h. The more satellites that orbit on the Figure-Eight orbit, the higher the elevation angles that are available. For example, a four-satellite system can provide elevation angles as high as 80 deg in Tokyo, and more than 70 deg nationwide. Modifications of the orbital parameters can give various different orbits with different service areas.

C. Deployable Antenna

To reduce the size of Earth terminals for very small aperture terminal (VSAT), maritime, land-mobile, aeronautical, and satellite phone users, one solution is to use higher frequencies. In such small-user to larger-Earth-station links, the C/N_0 or carrier power to noise density ratio in the uplink, is the main contributor to the total C/N_0 because the C/N_0 in downlink is much larger than the C/N_0 in uplink. If a satellite uses a higher-frequency band but the same size antenna, the gain of an onboard receiving antenna is proportional to the square of frequency, and the path loss is inversely proportional to the frequency squared. In total, both factors cancel each other. Because the antenna gain of a user terminal, and thus its EIRP, is proportional to the square of frequency, the total C/N_0 is approximately proportional to the square of the frequency if the noise temperature of the satellite receiver is not largely dependent on frequency.

The other solution to reduce terminal sizes is to increase the satellite's G/T, particularly by increasing the antenna size. Recent satellites for mobile and VSAT communication services are embarking larger-aperture and larger-gain antennas. For example, the Thuraya satellite has a single satellite antenna with a 12.25-m × 16-m mesh reflector, and the ACeS L-band multibeam satellite carries two 12-m-diam reflectors.

These large-aperture antennas typically have meshed rather than solid reflectors, which are folded for launch and deployed on orbit. There are two types of deployable antennas—a wrap-rib type and an umbrella type. The wrap-rib antenna consists of flat ribs that can spread out into radial directions, a center hub, and a reflective mesh that extends across the ribs. The umbrella-type deployable antenna also consists of multiple ribs and a center hub, but it can achieve expansion by having the ribs open up from their base like an umbrella. Figure 6.4 shows the ETS-VIII satellite with multiple umbrella-type reflectors. The unfolded reflector size is 19.2 m × 16.7 m. The reflector is assembled with 14 umbrella-type modules, each of which is deployed by using stepping motors. To form the multibeams, a phased-array antenna feed is used, consisting of 31 elements. The feed array is located approximately 1 m away from the focal point. Two separate reflectors for transmit and receive in S-band are installed on the satellite to avoid signal coupling and interference by passive intermodulation.

D. Active Phased-Array Antennas (APAA)

An array antenna can not only steer its beam direction but can also generate a particular radiation pattern by using a large number of radiating elements. To achieve a particular performance, a particular selection of radiating elements and array geometry, and the determination of the element excitations, are required. Several types of radiating elements are used in satellite antennas—horn, dipole, helix, and microstrip patch. Figure 6.5 shows an S-band phased-array antenna, which was installed in the ETS-VI satellite for intersatellite links between GEO and LEO. The size is 1.8 m × 1.8 m. The array consists of 19 radiating elements, each of which is equipped with one phase shifter for transmit beams and two phase shifters for receive beams so that each of these beams can be electrically scanned independently.

Figure 6.6 shows a Ka-band active phased-array antenna of the Japanese Gigabit Satellite. Two antennas with an aperture diameter of 2.2 m for transmit and 1.5 m for receive are installed on the same surface of the satellite's Earth panel. The APAA consists of 38 subarray units, each of which consists of 64 horn-antenna elements with a mutual spacing of 2.2λ. This figure shows this subarray unit. Four beams with scanning angles of 8 deg are available.

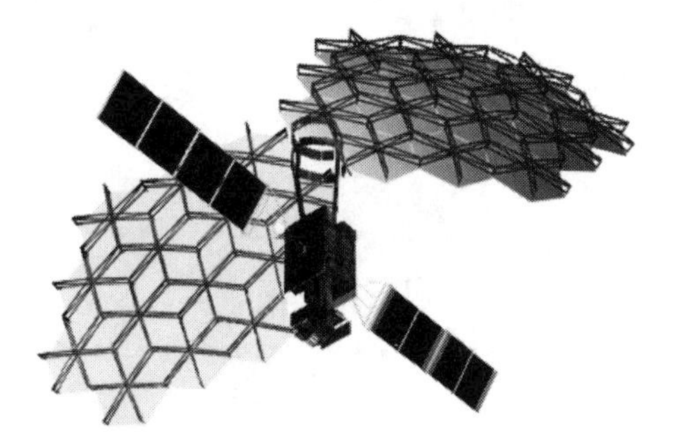

Fig. 6.4 ETS-VIII (Courtesy of NASDA).

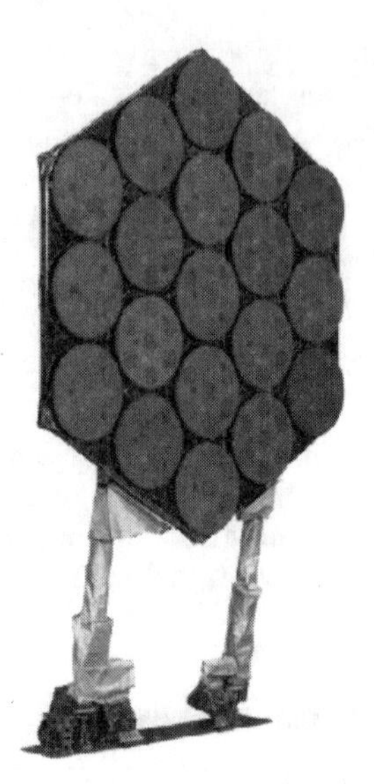

Fig. 6.5 S-band phased-array antenna of ETS-VI.

Fig. 6.6 Ka-band active phased-array antenna for Gigabit satellite.

E. Ultrawide Band[11]

UWB technology has the potential to provide high data rate communications and precise positioning under multipath environment. A UWB signal is defined as a signal that has the bandwidth larger than 25% of the signal frequency: $B > 0.25f_c$, where B is the bandwidth and f_c is the center frequency of the signal. Several types of UWB, such as pulse UWB (P-UWB) and frequency-hopping UWB (FH-UWB), have been proposed. P-UWB sends a pulse sequence, shown in Fig. 6.7, which occupies a frequency bandwidth of several GHz. FH-UWB is produced by hopping a spread spectrum signal over wide frequency bandwidth. Both methods can avoid the interference to other wireless systems by limiting the transmitting power, shaping the spectrum of pulses, and frequency hopping patterns.

In the United States, the FCC issued a ruling, in February 2002, that allowed UWB transmission in the band from 3.1 GHz to 10.6 GHz for communications and measurement systems. In Europe, the Conference of European Postal & Telecommunications (CEPT) is preparing to make European recommendations. In Japan, in August 2002, CRL started studies on UWB in microwave and millimeter wave for present regulation and future applications in cooperation with private companies and universities.

While achieving high data rates is a major R&D focus for communications, precise positioning is also attracting applications. The global positioning system (GPS) is providing high-accuracy position and time to an unlimited number of users in the world with relatively inexpensive devices. Because GPS satellite signals

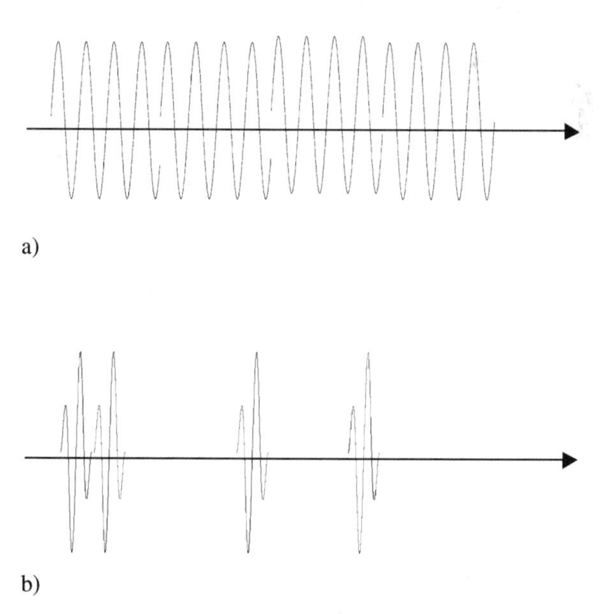

Fig. 6.7 UWB signal and carrier-modulated signal: a) Ordinarily carrier-modulated signal; b) P-UWB using band-limited monopulses.

are too weak to penetrate buildings and underground, satellite-based positioning services will have limitations in their use. GPS-based positioning services are not available in many places and times. Alternative positioning services are now being introduced using cellular-phone-based or other relatively narrow bandwidth positioning techniques. Indoor positioning will require higher positioning accuracy than that available out of doors with GPS. For example, the location of someone to an accuracy of better than 1 m may be required in some applications. UWB-based highly precise positioning will have a wide range of applications.

UWB is also a hot new topic as a new technology to improve personal wireless networks, but it is not clear whether it can be used in satellite communications. Studies will be necessary to use radio determination and broadband communications.

F. Optical Communications

When new large constellations of LEO and MEO systems are introduced, intersatellite links using RF and optical connections are expected to come into commercial operation. A number of observation satellites, such as the United States LANDSAT, French SPOT series, and Japan's MOS, JERS, and ADEOS, have been successfully operated to observe the Earth environment. Because these have high-definition images and observation data acquired from a lot of sensors, higher data rate links are required to send huge amounts of data to ground stations during the relatively short visibility periods from these Earth-orbiting satellites. Since the 1980s, optical satellite communications technologies have been investigated by researchers in the United States, Europe, and Japan. Figure 6.8 shows recent trends of such space optical links.

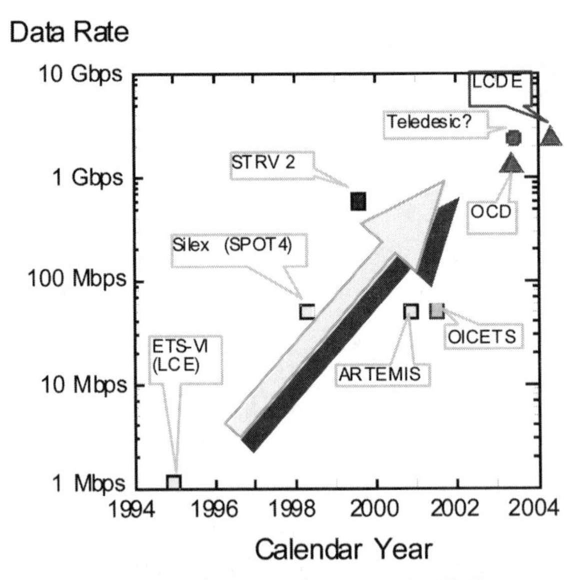

Fig. 6.8 Data rates of space optical links.

The United States has a history of more than 30 years of studying intersatellite laser communication technology, which has been carried out mainly by DoD and NASA/JPL. On 7 June 2000, TSX-5 (Air Force Tri-Service Experiments 5) developed by the Ballistic Missile Defense Organization (BMDO) was launched, and the first payload, named Space Technology Research Vehicle-2 (STRV-2), was used for optical communications experiments between the satellite and the ground station at a data rate of 1 Gbps. Because of unexpected large satellite attitude errors, the optical communications experiment could not be successfully carried out. In May 2001, MIT Lincoln Laboratory launched a geostationary satellite, GeoLITE (Geosynchronous Lightweight Technology Experiment), which embarks advanced optical communication systems.

Europe has developed a free-space optical communications system, named Semiconductor Laser Intersatellite Link Experiment (SILEX),[12] which consists of two optical communication payloads—one on the ESA ARTEMIS (Advanced Relay and Technology Mission Satellite) and the other on the French Earth-observation spacecraft SPOT-4. SPOT-4 was launched 24 March 1998. Optical alignment via internal calibration, the performance of the laser diodes, and programmed automatic acquisition and tracking of bright stars were checked. ARTEMIS was launched on 13 July 2001, but was put into a lower orbit than intended because of a malfunction in the upper stage of Ariane 5. However, recovery from the lower orbit to the geostationary orbit was carried out using the onboard electric propulsion system. On the way to the geostationary orbit, on November 2001, one-way optical communication experiments were successfully conducted from SPOT-4 to ARTEMIS at a data rate of 50 Mbps.

In Japan, NASDA started to study intersatellite laser communication systems in 1985, concentrating on the development of high data rate transmissions, light and small-size equipment, low-noise optical amplifiers, and acquisition and tracking methods. NASDA has developed its own laser terminal, laser utilizing communications equipment (LUCE) which is to be embarked on the Japanese OICETS satellite (Optical Inter-orbit Communications Engineering Test Satellite, see Ref. 13). The launch of OICETS was initially scheduled in 2001, but subsequently delayed. The OICETS weighs 550 kg, and will be launched into a circular orbit at an altitude of approximately 600 km and an inclination of 35 deg. The on-board laser terminal will be used for bidirectional optical communications experiments with ARTEMIS.

CRL developed laser communication equipment (LCE), which was installed on the ETS-VI satellite,[14] launched in 1994, and carried out the world's first optical communication experiments between the ETS-VI satellite and the optical ground station. In the downlink, a data transmission of 1.024 Mbps was carried out with a solid-state laser of 0.83 μm, and in the uplink an Argon laser with a power of 10 W and a wavelength of 0.51 μm was used. It was found that, in the satellite-ground links, a compensation technique against scintillation because of atmospheric turbulence is very important for reliable optical communications.

CRL and NASDA are now developing a laser terminal, which will be installed on the International Space Station, to conduct high data rate (2.5 GHz) transmission experiments. Japan's TAO is developing an intersatellite laser communication terminal for a broadband multimedia LEO system named NeLS (next-generation LEO system). In NeLS, 120 LEO satellites with 10 orbital planes

Fig. 6.9 AUJ-type antenna of NeLS satellite.

will orbit at an altitude of 1100 km. These satellites will be connected with intersatellite optical links of 1.55 μm in wavelength. Three types of optical antenna mechanisms were developed—Coude type, the active universal joint (AUJ) type, and the elbow type.[15,16] Figure 6.9 shows the AUJ-type antenna, which has an antenna aperture of 60 mm, an azimuth range of 360°, an elevation range of $\pm 20°$, an antenna height of 235 mm, and an antenna weight of 12 kg. To simplify the onboard receiver, IM-DD (intensity modulation/direct detection) with the Erbium doped fiber amplifier (EDFA) was selected.

G. Technological Challenges*

The advantages of intersatellite laser links over RF are higher-data rates possibly because of the higher frequency, lower divergence of the antenna beam, lower power consumption, and small and light terminals. The following are, however, technological challenges for optical satellite communications.

Higher data-rate intersatellite laser communication systems with small, compact, and light equipment produce a concentration of heat, and the design of onboard thermal control systems will be difficult because of a lack of space to radiate heat from the equipment surface. Large heat-emitting equipment and active heat control equipment will lose at least part of the advantages of lightweight and small size in the intersatellite laser communication systems.

Because of the narrow divergence angle of a laser beam and the finite speed of light, precise pointing and tracking systems are needed. Two-fold feedback-loop

*Private communication with Y. Arimoto, 2002.

tracking algorithms and precise tracking are used, but these mechanisms make optical equipment more complex, expensive, and less reliable because of the many elements required. Simple, low-cost, and highly reliable pointing, acquisition, and tracking mechanisms will be required.

Available frequencies for intersatellite laser links are four bands of 0.8 μm, 0.98 μm, 1.06 μm, and 1.55 μm. The 0.8-μm band has been used in onboard optical communication equipment of ETS-VI and OICETS, and this frequency is the optimum in quantum efficiency and sensitivity with silicon devices. Data rates as high as 1 Gbps will be available. The 0.98-μm band is suitable for low data rate communication links and for beacons for initial acquisition and tracking. The 1.06-μm band has been studied by ESA for coherent optical communication systems, and a high data rate for intersatellite laser links will be realized by using a solid-state Neodymium-Yttrium Aluminum Garnet (Nd-YAG) laser and homodyne PSK modulation. In the future, 1.55 μm is the most expected frequency band, because high data rates of up to approximately 10 Gbps and high sensitivity systems will be realized using EDFA.

V. Conclusions

The R&D targets for the next-generation communication satellites should be to aim for a space R&D program that affects other fields of science and technology through technology spinoffs, to provide cost-competitive solutions, and to promote collaborations with private sectors, governments, and academia. To realize the next-generation satellite systems, including Internet satellites, several advanced technologies will be needed, such as the use of higher-frequency bands (such as V/Q and W bands) light, deployable advanced antennas such as active phased arrays, ultrahigh-speed transponders, ultrasmall satellites, advanced Earth terminals, and improvements in positioning services, and orbital servicing. Several technologies such as HAPS, alternate means of orbit use, deployable antennas, active phased array antennas, UWB, and optical communications have been highlighted. It will be important to integrate a number of these advanced technologies in the design of the next-generation communication satellites.

References

[1]Pelton, J. N., and Bekey, I., "Advanced Geoplatform Concepts," *Proceedings of the TSMMW 2002 Conference*, 2002.

[2]Iida, T., and Suzuki, Y., "Satellite Communications R&D for Next 30 Years," *Proceedings of the 19th AIAA International Communications Satellite Systems Conference*, Toulouse, France, No. 233, April 2001; also, *Space Communications*, Vol. 17, No. 4, 2001, pp. 271–277.

[3]Pelton, J. N., "SCGII Delphi on Future of Satellite Communications," SCGII 2000, 2 June 2000.

[4]Pelton, J. N., "Telecommunications for the 21st Century," *Scientific American*, Vol. 278, No. 4, April 1998, pp. 80–85.

[5]Hase, Y., Wu, G., and Miura, R., "Wireless Communication Systems using Stratospheric Platforms," *Proceedings of the 18th AIAA International Communications Satellite Systems Conference*, AIAA, Reston, VA, 2000, pp. 251–258; also AIAA Paper 2000-1129.
[6]Miura, R., and Oodo, M., "R&D Program on Telecom and Broadcasting System Using High Altitude Platform Stations," *Journal of CRL*, Vol. 48, No. 4, Dec. 2001, pp. 33–48.
[7]Oodo, M., and Miura, R., "A Study of Frequency Sharing and Contribution to ITU for Wireless Communications Systems Using Stratospheric Platforms," *Journal of CRL*, Vol. 48, No. 4, Dec. 2001, pp. 33–48.
[8]*ITU-R Final Acts*, WRC-97, pp. 60/398, Geneva, 1997.
[9]Paynter, C., and Cuchanski, M., "System and Antenna Design Considerations for Highly Elliptical Orbits as Applied to the Proposed Archimedes Constellation," *Proceedings of the 4th International Mobile Satellite Conference*, Ottawa, 6–8 June 1995, pp. 236–241.
[10]Tanaka, M., Kimura, K., Kawase, S., and Wakana, H., "Applications of the Figure-8 Satellite System," *Space Communications*, Vol. 16, 2000, pp. 215–226.
[11]Darby, B., Diederich, P., Frazer, E., Harmer, D., and Morgan-Owen, G., "The Potential of UWB for Local Augmentations to GNSS Positioning," *Proceedings of the 20th AIAA International Communication Satellite Systems Conference and Exhibit*, AIAA, Reston, VA, 2002, pp. 1–10; also AIAA Paper 2002-2009.
[12]Sodnik, Z., and Lutz, H., "More than 20 Years of Laser Communication Technology Development in ESA," AIAA Paper 2000-1260, 2000, pp. 1031–1041.
[13]Suzuki, Y., Nakagawa, K., Jone, T., and Yamamoto, A., "Current Status of OICETS Laser-Communication-Terminal Development: Development of Laser Diodes and Sensors for OICETS Program," *Proceedings of the Society of Photo-Optical Instrumentation Engineers*, Vol. 2990, 1997, pp. 31–37.
[14]Arimoto, Y., Toyoshima, M., Toyoda, M., Takahashi, T., Shikatani, M., and Araki, K., "Preliminary Result on Laser Communication Experiment using ETS-VI," *Proceedings of the Society of Photo-Optical Instrumentation Engineers*, Vol. 2381, 1995, pp. 151–158.
[15]Suzuki, R., Sakurai, K., Nishiyama, I., and Yasuda, Y., "Current Status of NeLS Project: A Study of Global Multimedia Mobile Satellite Communications," *Proceedings of the 19th AIAA International Communication Satellite Systems Conference and Exhibit*, AIAA, Reston, VA, 2002; also AIAA Paper 2002-190.
[16]Suzuki, R., Nishiyama, I., Motoyoshi, S., Morikawa, E., and Yasuda, Y., "Current Status of NeLS Project: R&D of Global Multimedia Mobile Satellite Communications," *Proceedings of the 20th AIAA International Communication Satellite Systems Conference and Exhibit*, AIAA, Reston, VA, 2002; also AIAA Paper 2002-2038.

Questions for Discussion

1) **Calculate elevation angles of Molniya, Archimedes, and Figure-Eight satellites from a particular ground station.**

2) **In terms of intersatellite optical links between GEO and LEO when the distance between two satellite is 30,000 km, and the orbital velocity of LEO is approximately 8 km/s, what size point angle of GEO's laser beam is needed to track the motion of LEO.**

3) **Compare advantages and disadvantages of RF and optical links for intersatellite communications.**

4) **Find spinoff technologies from satellite communications technologies to other fields of science and technology.**

Chapter 7

The Next Thirty Years

Takashi Iida*
Communications Research Laboratory, Tokyo, Japan

and

Joseph N. Pelton†
George Washington University, Washington, D.C.

I. Introduction

IN THIS chapter, new, longer-term targets for research in the field of satellite communications during the next 30 years will be addressed. This analysis will focus primarily on geostationary equatorial orbit (GEO) platforms as the key type of future communications satellite system in terms of performance, coverage, and cost efficiency, although other types of systems will be addressed. Certainly, other technical innovations such as high-altitude platform systems or new types of satellite constellation may also evolve during the next 30 years and prove viable, but this chapter seeks to explore concepts by which, in particular, GEO systems and related technologies may achieve significant improvement in technical performance and breakthroughs in cost efficiency.

We will examine the reasons why satellite communication systems are needed in the information technology (IT) era, then address the need for longer-term satellite communications research and, particularly, the need for and role of governments in satellite R&D.

Many see the future vision of the communication satellite as increasing in capacity and performance in a similar manner to the recent rapid increase in capacity and performance of terrestrial links. There are many reasons why this might prove true. One can foresee the future of satellite communications as paralleling the growth in future generations of the Internet. Already, the most rapid drivers of fixed satellite systems are Internet-related applications, and this trend is expected to impact the future of broadcast and mobile satellite markets as well.

To respond to future market needs, three generations of Internet satellites are discussed and analyzed in terms of technology, cost, and technical performance.

---

*President.

†Director, Space and Advanced Communications Research Institute.

The first generation Internet satellite (namely, the 1G-satellite) could be launched in the 2000s, the second generation (namely, the 2G-satellite) in the 2010s, and the third generation (namely, the 3G-satellite) in the 2020s. The capacity of these satellites would steadily increase and the 3G would reach extremely high throughput levels in the range of several hundreds to several thousands of Gbps. As the discussion moves from the 1G-satellite to the 3G-satellite, we will be increasingly moving from the domain of engineering models to that of technical concepts. It is important, however, to note that there are a variety of new technical concepts that remain before us that can and likely will enable communications satellite systems to become ever more technically capable and cost efficient as we move into the 21st century.

Thus, the 1G-satellite is solidly grounded in today's research and is firmly based on recent broadband projects. The 2G-satellite includes much new yet reasonably well-confirmed technology to achieve more capacity and accommodate "regional to global" scales of communication. The 3G-satellite represents a much more advanced system and is in effect a "knee on the curve" of satellite research. This would likely involve the introduction of a future GEO platform, but of a much different type than those previously envisioned, which would have required very massive structures to be constructed in space.

A possible configuration for the 2G-satellite is discussed, which could come in two evolutionary variants. One would have more capacity and the other would be designed with more global service characteristics. Strategies to establish more capacity for this satellite design and the proposed configuration for the 2G-satellite are introduced. In addition, a proposed design for the global services satellite is analyzed and depicted.

Totally new types of GEO platform are discussed in the 3G-satellite section. In fact, three different concepts are presented as to how a GEO platform might be approached. This longer-term concept of satellite design and engineering provides insights into some dramatically new satellite system architectures, which could usher in a new age of dramatically more cost effective broadband satellite services. Although the concepts are now rather tightly conceived, a great deal of future design and engineering work is still needed to make these into viable systems that could actually be deployed and commercially operated in coming years.

The cost-competitiveness of the future communications satellite is a valid concern; thus, this topic is also discussed. This includes a discussion of the cost-competitiveness of the communications satellite over the various new generations of satellite design. Finally, we discuss the impact of future manned space development and its relationship to future satellite communications research and development.

II. The Necessity of Communication Satellites in the IT Era

A. The Necessity of Satellite Communications as a Complement to Terrestrial Networks

It is well known that the core competencies of satellite communication include wide coverage and access to remote and rural areas; broadcasting and mobile communications services; immediate installation and reconfiguration of communication links; wide and rapidly variable bandwidth; and disaster communication capability.

Recently, global terrestrial communications systems, represented by both fiber-optical cables and cellular mobile communications networks, have expanded greatly in coverage, performance, and cost efficiency. Thus, some people now question whether there is a viable role for satellite communications in the future. However, satellite communications, as we will show in this chapter, are still needed in many applications and circumstances.

1. Service to Remote Areas

Satellites are needed to complement terrestrial systems for two reasons. The first reason is that there are many areas that are too rural, remote, geographically isolated, or uneconomically viable for terrestrial systems to cover. Such problem areas exist in most developing countries, in ocean and arctic areas, and even in isolated parts of many of the developed countries. There is, for instance, approximately 25% of the U.S. and approximately 10% of the area of Japan, where terrestrial cellular mobile communications systems or broadband Internet are not readily available, even to small cities or towns. Satellite communication systems can easily resolve such problems because a core competence of satellite communications is wide-area coverage.

2. Service During Disasters

Satellite communications are needed particularly urgently to be able to take needed antidisaster measures. This lesson comes clearly from the Great Hanshin/Awaji earthquake (more commonly known as the Kobe earthquake). In this 1995 disaster, more than 6000 people died. Although this was the most dramatic instance, there have been many other cases where emergency reliance on satellites to provide disaster relief and communications has been critical. These instances include such events as the Tokai, Japan, flood of 2000, the Miyake Island, the Pele, and St. Helen's volcanic eruptions (in Japan, the Caribbean, and the United States, respectively), and many earthquakes in Mexico, Turkey, Russia, China, and India.

3. Wideband Internet

Satellite communications are critical to the expansion of the Internet worldwide. Internet capacity demand continues to increase rapidly, driven by the nearly doubling of the number of users each year for the last two years. More than 11% of world Internet service providers (ISP) already use satellites. The market for satellite Internet increased 8 times in these past 3 years.[1] Moreover, broadband satellites are expected to capture a substantial portion of the market for Internet access. Although estimates of the extent of this satellite market vary, they have ranged from 7 to 30% of the global market.[2]

Although the extent to which these trends will continue in the future is not certain, the need for satellites clearly remains. Even if the market share of future satellite communications decreases with regard to the total communication demands, as the capacity of terrestrial optical fibers increase, the demand for satellite communication itself will continue growing in the future. This is because communication demands overall are increasing tremendously.[3–5]

B. The Capability of Satellites to Cope with Capacity Demands

When the capabilities of satellites began to be overtaken by terrestrial optical fiber (in terms of relative throughput) in the beginning of the 1990s, the authors recognized that there would still remain a complementary need for both satellite and optical fiber to work together, because of rural and remote coverage, broadcast applications, and disaster requirements. The world's telecommunications and IT needs are sufficiently diverse that neither satellite nor optical-fiber networks alone could ever meet them. Given this longer-term need for high speed and large capacity, the concept of a "gigabit satellite" was studied by the authors as early as 1992 and formally proposed by them in 1993.[6] Since that time, the trans-Pacific high data rate satellite communication experiment in 2000 has demonstrated that satellite links have no problem in supporting high data rate transmissions and seamless connections with optical fiber via ATM (asynchronous transfer mode). Furthermore, the development of sophisticated "spoofing techniques" have verified that very high speed Internet/IP transmissions of 522 Mbps can be sustained with greater than 80% efficiency (see Chapter 3).[7]

There is, in general, an ongoing commercial success with direct broadcasting satellite services across the world, with examples such as NHK's DBS (Japan Broadcasting Corporation's Direct Broadcasting Satellite) in Japan, SES and Eutelsat in Europe, and DirecTV, Echostar (and soon Americom2Home) in the United States. By year-end 2005, there may be well over 100 million direct-to-home (DTH) satellite subscribers. In addition, new direct radio broadcast satellite networks are now being deployed and are meeting with modest but promising results. These include Sirius, XM Radio, WorldSpace, and others still to come.

Mobile communications as a market has been the most problematic in recent years because of the lack of sufficient customer demand to support the Iridium, Globalstar, ICO, and Orbcomm systems, which have all had to seek protection from their creditors by declaring bankruptcy. Networks such as Inmarsat, TMI, Thuraya, ACeS, and Garuda are, however, still financially viable and are seeking to establish greater market penetration and revenues for their services.

C. Necessity of R&D by Government

Currently, there are a number of strong governmentally funded satellite communications research programs around the world. These include those of the European Space Agency, as well as national space programs in Italy, Germany, and France. In Canada, the Communications Research Centre and the Canadian Space Agency have an ongoing program with guaranteed funding for a five-year program. In Japan, the Communications Research Laboratory, together with the Telecommunications Advancement Organization (TAO), the National Space Development Agency (NASDA), MITI, and other centers of excellence have a strong R&D program. Russia, China, India, Brazil, Israel, Korea, and the Ukraine also have space programs, with most of these programs giving primary focus to space applications and satellite communications.

NASA switched its satellite communications R&D policy in the late 1990s to reduce its commercial satellite research efforts and to focus instead on the communications needs of NASA's own programs, but there has been recent consideration to reviving commercial research programs in this area.[8]

In the defense sector, U.S. research progresses in the fields of high data rate communications and, especially, in optical communications. Thus, the necessity of R&D by governments in this field is broadly evident. China, India, Brazil, Korea, and Israel spend 30 to 40% of their space research budgets on space applications. Up to 10% or more of the space program's budgets of Japan, Canada, and Europe are directed toward commercial satellite applications. Thus, there is a general perception all over the world that governmental research programs are needed to continue to develop the technology seeds for tomorrow's communications satellite programs.[9]

For similar reasons, governmental research programs should also continue their research programs to develop new IT infrastructure technology and to develop technology and new standards to help maintain the complementary and seamless interconnectivity of satellite systems with terrestrial systems.[10] Finally, in today's complex and sometimes threatening international political climate, government research to maintain the reliability, system availability, and viability of satellite networks against terrorist attacks or even interference or infiltration by criminal interests needs to be a part of the overall R&D program.[10]

Governments can conduct long-term R&D as best fits their own economic and political system and also can involve industry and academic institutions as appropriate to their national needs. Nevertheless, most space-faring nations will conduct some R&D programs, and in commercial networks, a high degree of international collaboration will continue. Clearly, there is a difference in satellite communications R&D as conducted by government and that carried out by private enterprise. Finding the best formula for this interaction will be key to the future of the industry, particularly in the development of the most advanced technology described in the remainder of this chapter. Part of this program of collaboration will involve the issue of flight demonstrations or in-orbit testing of the latest technology. Flight demonstrations are necessary for technology to be usable in private enterprises, and it is in this area where the formula for collaboration between and among government, industry, and academia is most critical.

Most governments thus perceive the need to maintain research initiatives in telecommunications and IT technology to pursue longer-term programs of national interest. This means that governments must be able to attract outstanding telecommunications, computer, and IT researchers and engineers, while not, however, competing unduly with the needs and interests of industry.

For a number of reasons, international cooperation is necessary, and this is particularly relevant when it comes to satellite communications, which are broad in their coverage and involve international applications from the Internet to tele-education and tele-health services. In this respect, Japan, the United States, Canada, Europe, and Australia, along with other countries, have promised to support special undertakings. In the case of Japan, this includes an obligation to promote future satellite capabilities in the Asia-Pacific area and to contribute to bridging the "digital divide" (G8-Okinawa charter.)[11] Of course, the relationship between a developed nation and a developing country should be focused and contribute to education, health, democratization, and world peace by the circulation of information. Thus the perspective of antidisaster measures is indispensable to this overall process. The government should also construct the communication infrastructure, including urgent communication, to meet the previously mentioned values and needs. A number of countries have thus undertaken to include important social, education, health, and development goals in their future satellite communications research programs.

The R&D of satellite communication technology includes not only technology itself but also the expectation of spinoffs and/or stimulation to other technology and social applications. Finally, there might also be particular technologies that cannot be developed without investments too large to be sustained by a single commercial organization. It is in just such a circumstance where the R&D programs of governments can and do often hold special significance. As the scale of the technology increases in the next decades, this may become even more relevant than it is today.

III. A Future Vision of Satellite Communications

A. Three-Generation Approach to the Future of Communications Satellites[12]

To develop an infrastructure that operates in harmony with terrestrial communications systems and can provide complementary technical performance, the speed of satellite communications must meet or at least approximate the speed of the terrestrial communications system. This is quite challenging because there have been spectacular gains in fiber optic technology in the past few years. However, for the longer term, the speed of satellite communications must be substantially increased.

Therefore, we foresee both a demand for higher and higher throughput and for larger capacity satellite communications systems. It is from this perspective that it is useful to consider, as a representative case, the development over time of satellite systems for high-speed Internet access. Thus, let us consider communications satellite R&D over a long-term span of the next 30 years, separately addressing, the 2000s, the 2010s, and the 2020s as proposed and shown in Table 7.1.[12]

An example of a concrete project of the 1G-satellite is the design that was identified previously as the AAPTS (Advanced Asia-Pacific Telecommunication Satellite).[6] The AAPTS considered the service of providing 1-Gbps links to the Asia-Pacific region, using a very high capacity optical Inter-Satellite Link (ISL). This concept has been expanded to the Japanese WINDS (Wideband InterNetworking Engineering Test and Demonstration Satellite) project.[13]

The 2G-satellite will transcend the WINDS design and have a design concept that allows more capacity and can provide a more global service capability. The 2G-satellite design concept will also include the possibility of incorporating a next-generation low Earth orbit (LEO)[14] and global ring satellite.[15]

Finally, the 3G-satellite will need to introduce the concept of a GEO platform. In addition, human space activity beyond 2020 will expand outside of Earth orbit, extending, for example, to manned installations on the Moon. Thus, we should consider an appropriate communications system that can meet such requirements.

Here, four supplementary comments should be pointed out concerning the proposals in Table 7.1. The first is that a generation change every 10 years may be

Table 7.1 Proposed communication satellite (Ref. 12)

Timeframe	Generation of Satellite	Capacity*	Characteristic
2000–2010	First-generation Internet satellite (1G-satellite)	5–50 Gbps	AAPTS, WINDS, iPSTAR, etc.
2010–2020	Second-generation Internet satellite (2G-satellite)	50–500 Gbps	More capacity and (in contrast) more global service characteristic
2020–2030	Third-generation Internet satellite (3G-satellite)	0.5–5 Tbps	New types of GEO platform architecture and expansion of human activity

*See Section III.B.

considered to be too long in a time when rapid innovation is occurring. However, on the other hand, the systems that are being developed to have higher throughput and capacity also involve substantial investments of capital that imply the need for longer-term amortization. This indeed is a problem currently being faced by the investors in very large capacity fiber networks that have not attracted sufficient new market demand to support their previous rapid expansion. In the satellite industry, during the past two decades, satellite generations have moved generally in what might be called "dog years," or seven-year time periods. For the future, where large jumps in capacity and throughput must be envisioned, while a generation of 10 years may be considered to be quite long, it nevertheless must be reasonable because of the longer period of amortization required for each generation of infrastructure.

The second consideration is the broader service category of Internet satellite. Because of digital multiplexing and the expanding applications of IP services, this new type of satellite may be considered equipped to provide fixed, broadcast, and mobile satellite communications services.

The third comment is that regional or domestic service categories and networks could give way to more globally-oriented services and networks. By the time a 2G-satellite design is developed, small-scale and geographically focused networks could well prove to be inefficient from both a service and economic perspective. However, it must also be noted that political, national trade, and other considerations could lead to results other than the most efficient technical design.

In the second generation, therefore, global reach (including local, national, and regional) capability could well be required. In this case, it may be preferable to design and deploy a "ring satellite" system and GEO satellite connections with LEO satellites (or even high-altitude platform systems). This may hinge not only on service needs, system economics, and trade considerations, but also on issues related to satellite transmission latency.

The fourth consideration is that it may not be economically feasible to launch satellites such as the 2G-satellites frequently, because they are of such a large size (and therefore expensive). However, it is also important to harmonize the development and flight of space-related technology with the development speed of IT technology. Small satellites incorporating technology improvements could be launched at much more frequent intervals. A small satellite program could also take advantage of this frequent launch sequence by including space infrastructure experiments, positioning technology, and new LEO technology. This issue will be discussed in more detail later in this chapter.

B. Target of Satellite Capacity

We would now like to consider the capacity of the proposed future generations of satellites. The capacity of terrestrial communication systems increases tremendously year by year, principally by the expansion of the technology of wavelength division multiplexing (WDM). A prediction based on past trends in this area is shown in Fig. 7.1. The terrestrial access network part, in the 2000–2015 timeframe is based on Ref. 16. For the backbone network, the dotted thick bold line shows an extrapolation by the authors, considering that there could be a physical limitation in the capacity of a single optical fiber cable, for example a few petabits per second. It is assumed that the capacity increase has almost the same trend for the backbone network as for the access line network.

According to Fig. 7.1, the capacity of a single fiber as a backbone network, shown in the upper part of Fig. 7.1, is roughly between 350 Gbps and 1 Tbps in the 2000–2005 timeframe, will increase to at least 2.5 Tbps in approximately 2010, and potentially to even much higher levels (again it is assumed that market demand will control these rates of increase as much as the available technology). On the other hand, the capacity of fixed-access links is between 56 kbps and 10 Mbps in the 2000–2005 timeframe, and will increase to as much as 2.5 Gbps in approximately 2015. The mobile access is between 10 kbps and 2 Mbps in the 2000–2005 timeframe and will increase to between 2 and 30 Mbps in approximately 2015.

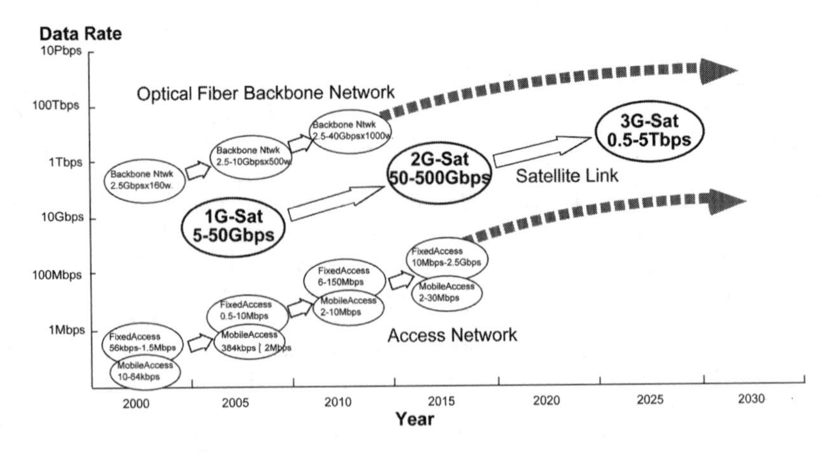

Fig. 7.1 Data rate prediction for various transmission media.

Satellite capacity should be increased in harmony with the terrestrial capacity, although complete parity currently seems impossible. Satellites can be used not only as a backbone network but also as a fixed- or mobile-access network. Therefore, we feel safe to assume here that the satellites will share at least 1% of backbone network capacity.

The gigabit satellite (or WINDS satellite) has 1.2 Gbps/channel and its total capacity is estimated to be approximately 6 Gbps. On the other hand, the iPSTAR satellite, to be launched in 2005, is already planned to have a total of 50 Gbps capacity.[17] Therefore, it is reasonable to assume that several to several tens of Gbps (5–50 Gbps) is the total capacity of the 1G-satellite. (The development of new and more spectrum-efficient modems that can transmit at levels of 4 bits/Hz or more could boost these figures up to substantially higher levels.)

As far as the 2G-satellite is concerned, we have assumed that its capacity should be at least 10 times that of the first generation. Therefore, a total capacity per satellite of several tens to several hundreds of Gbps (50–500 Gbps) for the 2G-satellite would be the design goal. Finally, it is further assumed that the goal for the 3G-satellite would be 10 times the capacity of the 2G-satellite. This then would be a goal of several hundreds to several thousands of Gbps capacity (500 Gbps–5 Tbps). Although these capacity targets might seem too high to be realized, it should be noted that some of these gains might come from modem development rather than satellite design gains. The bottom line is that generally comparative values appear needed for satellites to survive as a viable media in the future IT environment.

Regardless of whether our goals are overly optimistic or not, what is clear is that the "required" capacity of the 3G-satellite cannot be realized by a conventional type of satellite. Therefore, something along the lines of an advanced GEO platform design would seem necessary, and some new technical breakthroughs are needed.

Clearly, the capacity targets set forward in this section are only rough order of magnitude (ROM) estimates, but they are, nevertheless, not unreasonable R&D goals. We believe, on the basis of current trends in satellite and terrestrial technology, that something like an advanced GEO platform is needed to proceed from the 2G to the 3G era.

IV. Consideration of a 2G-Satellite

This satellite design evolves from reviewing technology trends in the communications satellite field and assuming the specification of the satellite needed to match the projected throughput, capacity, and cost-efficiency goals.

A. The 2G-Satellite with More Capacity[18]

1. Technology Trends for Communications Satellites

Certain overall technology trends for communications satellites are shown in Table 7.2. The capacity of a communications satellite is highly dependent on its ability to generate electric power. Most of the present commercial communications satellites are in the 8- to 10-kW class, and the current maximum is approximately 15 kW. In the future, however, satellite manufacturing companies in both Europe

Table 7.2 Trend of commercial satellites

Item	Trend
Generation power	8 kW → 15 kW → 20 kW*
Frequency band	Shift to the Ku (14/12 GHz) band or higher
Type of transponder	Simple bent-pipe type will remain the mainstream
Power amplifier	Embarking 100–150 TWTAs of 100-W class
Type of antenna	Reflector types will remain the mainstream for several years, with satellites embarking 4 or more reflectors of 2 m class diameter for FSS and BSS services, and one or two unfurlable antennas of much larger diameter for L/S-band mobile services.

*The satellite bus with 20-kW-class power generation is planned to be developed.

and the United States propose to develop satellite buses in the 20-kW and above class.

C-band and Ku-band satellites are still the mainstream at present. However Ku-band predominates in most of the satellites being ordered today, and Ka-band is expected to become more highly used in the future. Ka-band is typically planned to satellite Internet in the relatively near future, and ultimately the Q/V-band will probably follow soon after. What is most in question is whether simple bent-pipe type transponders, because of their economy and flexibility for changes, will remain the mainstream in the future. There are various other factors that could point toward increased use of onboard processing and signal regeneration. These factors include a move, driven by a need to obtain more usable spectrum, that leads to increase the number of spot beams and to the need to rapidly interconnect smaller and smaller sized beams; the need for signal regeneration to cope with rain attenuation problems; and the increased speed of an onboard processor that operates at ever-decreasing costs. These and other factors could point toward more onboard processing and signal regeneration by “smart transponders” in the future, at least for some services.

For power amplification and transmission, the use of traveling wave tube amplifiers (TWTA) is a mainstream at present, and this trend probably will not change, at least from the viewpoint of electrical power efficiency. Solid state power amplifiers (SSPA) have become more and more disadvantageous from the electrical power efficiency point of view as the frequency bands shift to higher and higher levels, but are clearly preferable when it comes to lifetime. The power efficiency issue is thus expected to become a key research focus in the future.

In addition, the performance of TWTAs has recently improved in both output power and reliability. Their in-orbit operating lifetimes have been expanded to approximately 20 years, while the output power has been increased to more than 100 W in the Ka-band. This issue of TWTA vs SSPA appears to be another key area for satellite research programs.

The link budget for satellite communications, especially for GEO satellites, is very demanding. Thus, it is difficult today to design satellite modems that can transmit with much greater efficiency than 1 bit/Hz. Nevertheless there are many research programs that are seeking to devise exceptionally "smart modems" that might be able to improve transmission efficiency. To the extent that new modem technology could make this possible, this could represent a substantial gain for the satellite industry. It would make a much greater impact on satellite transmission systems, which today operate inherently at a disadvantage compared to the very low noise and near-zero, bit error rate (BER) environment enjoyed by fiber optic networks.

For fixed and broadcast services, multibeam reflector antennas are expected to be the mainstream in the antenna field for the time being. A tendency to carry multiple reflector antennas (more than four) of the 2-m class will be continued except for special cases where phased-array antennas may be used. For mobile services, which are expected to remain predominately based on the use of GEO satellites, multibeam systems using unfurlable/deployable reflector antennas of up to 50 m in diameter can be envisaged.

2. *Assumed Specification*

A large-sized satellite is assumed from the recent technology development status point of view, as previously mentioned. As far as the size of the satellite is concerned, the weight of the satellite is determined by the capability of the launcher, which ultimately determines the size of a satellite. For example, an ESA Ariane 5 can accommodate satellite payloads up to 4.5 m in diameter with a mass of up to 6.8 tons [for a geostationary transfer orbit mission, (GTO)]. The NASDA H-2A launcher will accommodate payloads of comparable diameter and also has the ability to launch a satellite of more than 6 tons into GTO. In satellite lifetimes with such weight restrictions, the use of ion engines and lithium ion batteries are assumed. The assumed characteristics of a large-sized 2G-satellite are shown in Table 7.3.

As far as the capacity is concerned, the antenna beam allocation and services are not considered; only the capacity of the satellite is considered here. A 6.8-ton-class satellite can embark 100 transponders (in fact, the current Intelsat 9 satellite already hosts some 100 transponders, but these are of much lower throughput design). A high-capacity transponder of 622–1244 Mbps has now been developed in the WINDS project.[13] Therefore, the total capacity of the satellite might be 60–120 Gbps. We assumed that the capacity of the 2G-satellite would lie in the range of 50–500 Gbps. Therefore, capacity of 60–120 Gbps places this satellite at the low end of our assumed requirements range. It is reasonable, therefore, that we consider this satellite only as the first step toward the ultimate goal of the 2G-satellite design, and will refer to it herein as $2G_{1S}$.

3. *Configuration of a $2G_{1S}$-Satellite*

Given that the number of TWTAs is assumed to be approximately 100 in the budget of an assumed $2G_{1S}$-satellite, there are then the following problems to realize such mission equipment:

1) A large footprint to embark the mission equipment is required it is estimated that more than 45 m^2 is necessary).

Table 7.3 Model case of $2G_{1S}$-satellites

Item	Contents	Remarks
Number of transponders	100	100-W-class of output power
Maximum capacity	60–120 Gbps	622–1244 Mbps per transponder
Number of antennas	4	2 m class of diameter
Consumption power	18 kW	
Generation power	22 kW	
Weight of onboard equipment	1100 kg	
Weight of bus	2600 kg	
Weight (dry)	3700 kg	
Weight of thruster	3100 kg	
Launch weight	6800 kg	
Auxiliary thruster	Ionized engine	
Heat management	Deployable radiator	
Battery	Lithium ion	

2) The design of heat dissipation structures becomes critical because this becomes more important when the high power equipment mounting footprint requirements increase.

The position of thruster tanks can interfere with the desired footprint expansion when a central cylinder type of bus structure is adopted.

A system of separating the mission equipment part from the satellite bus part is possible and would serve to help solve the previously mentioned problem. To obtain a large equipment mounting footprint, the mission equipment (mission module) would configure a part of the satellite bus by combining vertical or horizontal panels together. The thruster tank and oxidant tank are then put in the lower part of the satellite bus. The heat transfer system, as an integral part of the panel system, would provide direct cooling of the TWTAs. This would consist of a flexible loop heat pipe (FLHP) and a deployable radiator.

An external view of a typical $2G_{1S}$-satellite configuration is shown in Fig. 7.2. As previously mentioned, an increase of satellite power generation is a critical design feature needed to accommodate the large capacity and high throughput rate. To increase the power of the onboard solar array, a two-dimensional deployable solar paddle could provide superior power generation capacities for a $2G_{1S}$-satellite.

4. Technology for a 2G-Satellite

A bent-pipe type transponder, TWTAs, and a reflector-type antenna are the mainstream in the present commercial communications satellite. However, TWTAs are bulky and require a lot of volume in relation to an SSPA. Thus, there is a practical limit to the number of onboard transponders. As far as the power amplifier is concerned, an SSPA could be developed instead of a TWTA. The output power

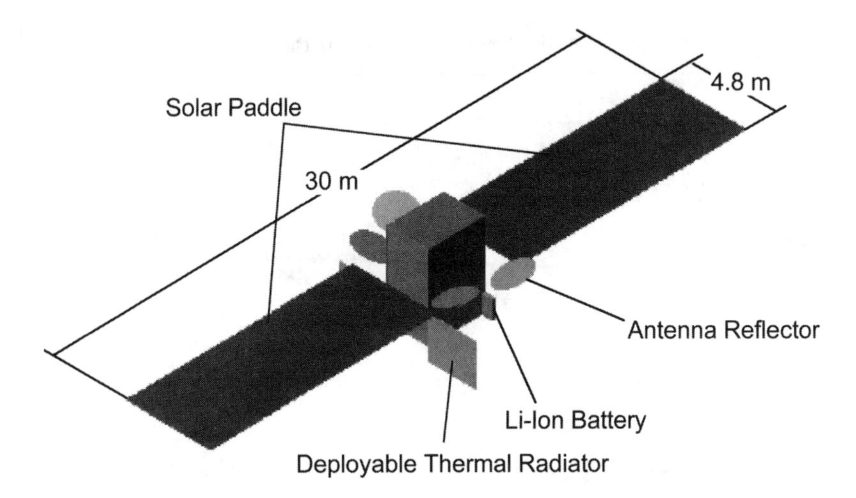

Fig. 7.2 Configuration of a $2G_{1S}$-satellite (deployed status).

capability and the efficiency at present in the C-band for an SSPA are comparable to those for a TWTA. A larger number of transponders will be able to be embarked in a limited satellite space in the future if high-powered SSPAs are developed in the Ku- and Ka-bands. Development of wideband gap devices (WBG) using Gallium Nitride (GaN) is being researched as a way for realizing such a high-efficiency and high-powered SSPA. This device might be able to provide an innovative way to realize both high output power and high efficiency.

Furthermore, in some R&D satellites, systems that use an active phased-array antenna (APAA) and an onboard processor (OBP) are under study. For the APAA, an SSPA is typically used, but the output power is low. This is the decisive reason that an SSPA at these higher frequency bands is not used commercially, because its power efficiency is significantly inferior to that of a TWTA. In addition, although OBPs have been developed to support voice communications, this is a relatively low-speed application from the viewpoint of the processing ability. OBPs using onboard ATM switching and IP routing are already in an advanced state of development. Thus, all of the critical technologies for future multimedia communication satellites seem to be on track to be available when they are needed.

In a 2G-satellite, a breakthrough in terms of improved satellite modems, coupled with improvements in SSPA electric power efficiency, plus the new signal processing ability of OBP, are critical technologies. Table 7.4 shows a trend of key technologies of the future mission equipment needed for the 2G-satellite.

B. Expansion to Global Services

A satellite communication system naturally lends itself to being able to support global communication services, but there is no true global system today using intersatellite links (ISL) except for military systems. In this subsection, the idea of constructing a gigabit information "space corridor" is discussed as an example of a global satellite communications system using ISL.

Table 7.4 Trend of key technologies for the future satellite

Items	Present	Future
Type of transponder	Bent-pipe	Regenerative
Power amplifier	TWTA	SSPA (WBG device)
Type of antenna	Reflector	APAA
Type of feed	Single-horn-feed	Digital beam forming feed/ optical feed
Onboard processor	N/A	High-speed OBP/ATM processor
Modulator/demodulator	N/A	Advanced "intelligent" software MODEM

First, we imagine a "satellite ring" of satellites in non-GEO equatorial orbits connected via ISL, and high data rate (HDR) space-to-Earth communications links to be able to establish an HDR link to a specific point on Earth at any time. Two types of equatorial orbit are considered—GEO and non-GEO. The GEO satellites are configured according to the design envisaged for the AAPTS. In this case, the merit of the ISL is to reduce the propagation delay between two Earth station locations, for example, to keep the delay from New York to Kuala Lumpur within 400 ms. To reduce the delay time more, it is necessary to use non-GEO satellites, preferably in the equatorial plane.

An example of the non-GEO system described here is a system called Space Information Super Corridor on an Equatorial Orbit (SISCEO),[15] whose specification is the following

a) Orbit: Circular orbit with zero inclination (Equatorial orbit)
b) Altitude: 15,500 km
c) Number of satellites: 6
d) Data transmission rate: Both ways at 10 Gbps
e) ISL configuration: Optical ring links in both directions
f) Space to/from Earth link: Optical or millimeter wave
g) Minimum elevation angle for the Earth station located at around 40 deg. North or South: 30 deg.

The configuration of the SISCEO is shown in Fig. 7.3. From the viewpoint of decreasing the propagation delay, in this case the lower-than-GEO altitude of 15,500 km was selected, for which case the number of satellites needed would be six. One could, of course, increase the number of relay satellites and lower the altitude, or decrease the number of relay satellites and raise the altitude, but the SISCEO system is considered herein as a reasonable approach. In this system, the elevation angle would typically be between 30 and 42 deg. In-orbit tests would be needed to verify the adequacy of these elevation angles.

The ISL configures the ring communications links surrounding the Earth. This idea is almost the same as one described in Ref. 15. One of the rings is used for the link in the clockwise direction, while the other is for the counterclockwise direction. These links can be configured as a redundant communications link,

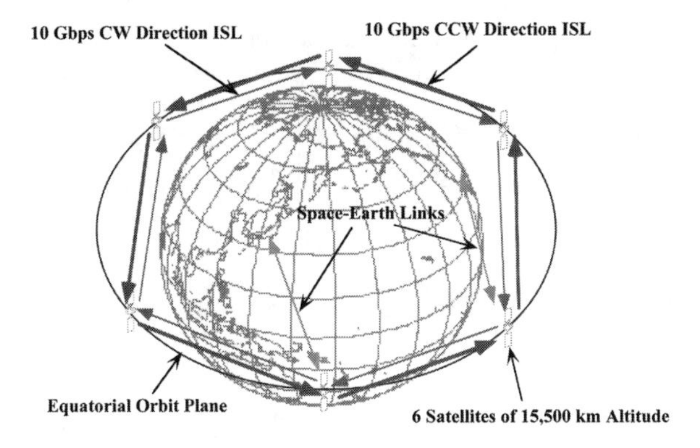

Fig. 7.3 Configuration of the SISCEO system.

although the propagation delay for a round trip is approximately 438 ms (131,268 km). The key to such a future satellite ring configuration has probably more to do with economics, markets, and charging policies than it does with the technology. The charging mechanisms by which satellite networks would use such networks and how the cost of this infrastructure would be paid for over time is probably the biggest challenge. In some ways this is like the classic "economic problem" associated with the creation and operation of a lighthouse, with the need to determine how the various ships at sea would pay for their usage of this facility.

V. Three New Designs For Advanced GEO Platform Systems as 3G-Satellites[19]

The major alternatives for the future of the satellite industry are continued migration to higher and higher frequencies, or the development of new satellite architectures such as advanced GEO platform systems that allow for dramatically increased levels of frequency reuse via geographic isolation of narrow spot beams. Totally new designs that use new materials and new concepts in phased-array antennas can provide a pathway to significantly more cost-effective satellite networks and can also facilitate new types of user microterminals. This section sets forth three different concepts of how GEO platforms might be approached. This involves speculation about designing and engineering some dramatically new satellite system architectures that could usher in a new age of more cost-effective broadband satellite services. A further benefit of such dramatic new designs is that the very high power beams emanating from advanced GEO platforms could also work more effectively to handheld or even wrist-mounted user microterminals.

An ongoing study conducted during the last two years by the Space and Advanced Communications Research Institute at George Washington University (GWU), on behalf of the Communications Research Laboratory (CRL) of Japan, has explored three different models of future satellite systems that would use breakthrough technologies to meet new satellite applications and meet new market demands with breakthrough economies of service.[20] The study only gives preliminary consideration to new broadband service demands for such satellite systems, but mobile broadband Internet, new air traffic monitoring systems, and new types of scientific research applications could reasonably demand such satellite service capabilities within some 15 to 20 years from now. These systems are described on the basis of progressively advanced technologies that can deliver increasingly higher throughput performance and improved cost efficiency.

A. Concept 3-1: Tethered GEO Platform Network

This first conceptual model of a future GEO platform system, as shown in Fig. 7.4, represents a logical next step in the extension of what might be called conventional satellite technology. The GEO platform concept will nevertheless still be able to realize dramatic performance gains. The key aspect of this new design is the use of new low-mass but high-performance polyimide materials to form new, very-large-aperture parabolic antennas. This new model would also exploit new tether technology and implement new highly efficient solar cell technology. These three factors, plus high-efficiency satellite modems and other less critical technologies can allow for the design of new GEO platform systems that are considerably more cost effective than today's satellites. The specific innovations would be as follows

1) Large-scale, large-aperture, multibeam antennas that use polyimide materials rather than metal, carbon epoxy, or deployable truss antennas. The net savings in antenna mass to effective gain performance could be very significant.)

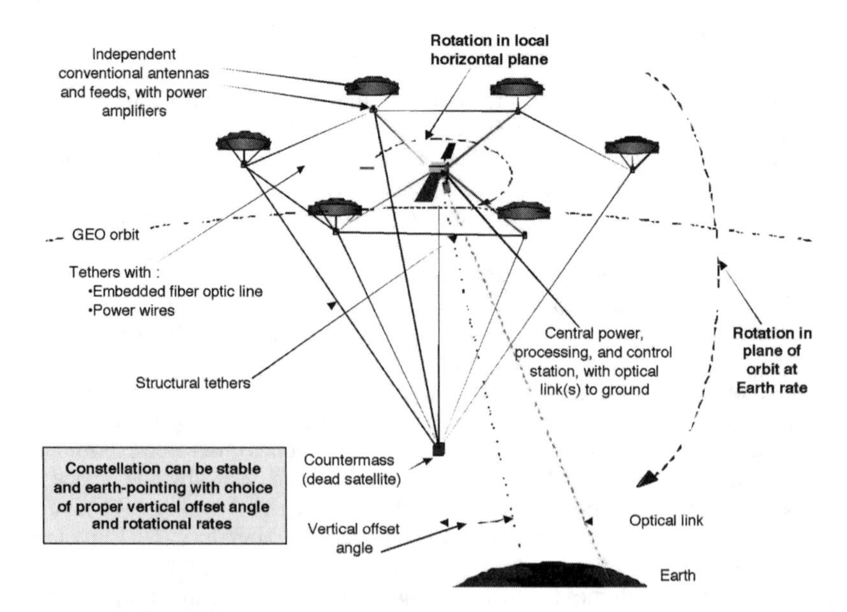

Fig. 7.4 Tethered GEO platform network. (Published with permission of Joseph N. Pelton and Ivan Bekey.)

2) High efficiency new "rainbow solar cells" that could allow net conversion efficiencies up to 65%. It is possible that lightweight and low-mass "lenses" could also be deployed to allow solar cell systems to see a high number of "multiple" suns.
3) Use of low-mass, high-tensile-strength tether systems to allow deployment of several GEO platform antenna systems as a combined network sharing a common power system.

In the configuration shown in Fig. 7.4, six very-large-aperture, multibeam antennas are assumed, although such a configuration could also be considered if one were to deploy a lesser number of such antennas. In ROM figures, each parabolic antenna might be some 25 to 50 m in aperture diameter, and each antenna might operate at up to approximately 150 Gbps (the total GEO platform would thus constitute 900 beams of 1 Gbps each). The net mass in GEO orbit might be approximately 15 metric tons and the system lifetime would be some 20 years. The estimated fully deployed system would be, in ROM figures, approximately $4.5 billion. The development of new, larger, and more cost effective launch vehicles that would allow these systems to be launched as an integrated system, rather than having to be deployed and assembled in space, would represent a major cost saving.

B. Concept 3-2: Advanced Piezo-Electric Shaped Antennas GEO Platform System

This GEO platform satellite system is based on the formation of shaped piezo-electric antenna structures. These structures must be exposed to continuous radiation to maintain their shape, although if a flat-shaped, phased-array antenna is used, this requirement is somewhat less exacting. In any event, in Concept 3-2-A (see Fig. 7.5) or 3-2-B (see Fig. 7.6), the piezo-electric antenna structure must be radiated from the location of the core feed structure to maintain the required shape as shown in Fig. 7.5.

If one were to deploy a parabolic-shaped reflector, the membrane would be constructed as a sandwich of two bonded piezo-electric materials, polarized in orthogonal directions, and constituting a bimorph. An electron beam would scan across the entire surface, with the charge being deposited at any point determining the local bending of the surface at that point. The beam current and the back electrode voltage are appropriately modulated such that a desired charge distribution is deposited within the membrane.

The beam and back voltage are made responsive to a separated optical figure sensor that determines the actual membrane surface and compares it to the reference (or desired) surface shape, and generates correction commands that are sent to the electron beam generator. The result is that the surface shape is forced to conform to the reference shape at all times, resulting in a closed-loop surface shaping and correction system.

This system thus could take an initially shapeless membrane, cause it to assume a parabolic desired figure, remove deployment wrinkles, and maintain that figure despite internal or external disturbances. Because the membrane is very thin and can be tightly folded (because the system can remove folding wrinkles), it can be

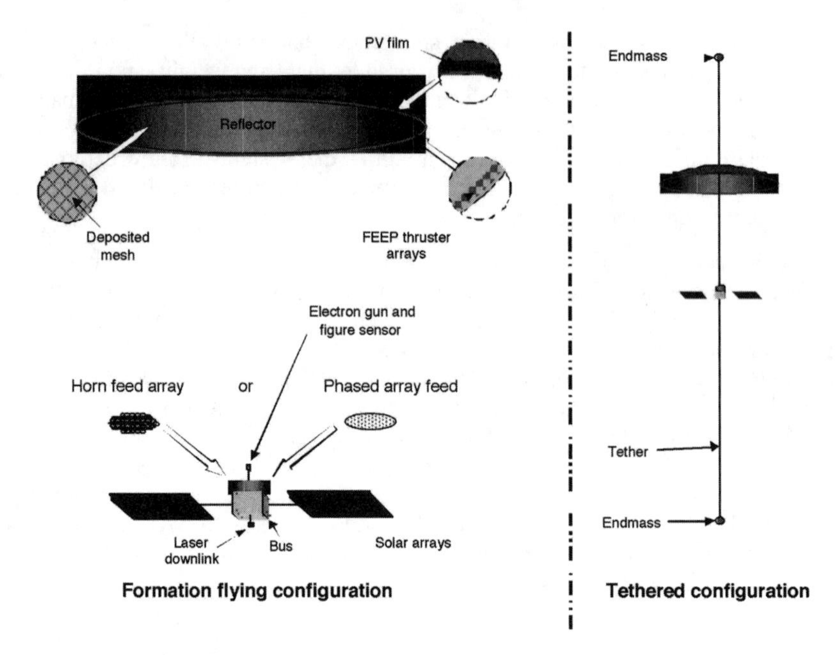

Fig. 7.5 Advanced piezo-electric shaped antenna GEO platform system in either free flying or tethered configuration (Concept 3-2).

tightly packaged. This means that an extremely large aperture can be flown in a small volume launch vehicle shroud. In the timeframe that such a system might be deployed, however, a flat phased-array antenna design seems more likely.

The same principle, as applied to array-fed reflector antennas, could also be applicable to space-fed lens antennas. In this case, the lens consists of phase-shifting elements located on the surface of an adaptive membrane, whose shape is set and maintained flat by electron beam control responsive to an optical figure sensor. As in all space-fed lens antennas, the phase shift (or time delay for greater bandwidth) of elements across the lens is varied so that all paths from the ground to the feed structure are the same electrical length. Thus, all elements contribute signals that add coherently. The application to GEO Platform antennas is illustrated in Fig. 7.5.

Alternative versions of this approach to a large-scale platform would be for the lens and the separated feed structure (and bus) to be "flown in position," while on the other hand, the configuration could be maintained by means of a tether that holds the components together against the secular forces and thus minimizes propellant expenditures. The lens structure is the adaptive membrane, on which the active elements are mounted. These would be spaced nominally one wavelength apart. The periphery of the membrane is covered with microthrusters and thin film photovoltaic material for powering them and the elements. There is no structure and the membrane is not tensioned, though a variant with light tensioning is possible and future studies will undoubtedly seek the optimum design for such a system.

A lens is inherently far more tolerant of positional inaccuracies of the elements. An advantage, therefore, of using a lens over a reflector is that the surface accuracy of the membrane can be much poorer and the structure could, in theory, be much larger. The feed structure contains either a phased array to generate the separate beams or an array of horns or equivalent RF structures to generate the equivalent beam patterns and structure. While the space-fed lens is more tolerant of surface positional errors, it is more complicated, heavier, and more expensive than a reflector because of the very large number of elements required and their power consumption. As an example, a 30-m-diam lens at C-band might require up to 200,000 active phase shifting elements, each with input and output dipoles. Again, future studies will likely find a way to reduce the number of these elements down to a more reasonable number.

The approach (that is, of using piezo-electric materials) would allow a much larger antenna structure than other architectures that, together with OBP and multibeam phased-array antenna technology, could produce a very large number of spot beams of very high power and extremely broadband throughput capability. This design could also benefit from the "rainbow" solar cells and tether technology similar to that envisioned for Concept 3-1. Overall, the net efficiency gains could be even larger than those of Concept 3-1. Future design work will determine whether it is best to use a conventional parabolic antenna design, or to use a high-efficiency, phased-array antenna.

In ROM figures, one might envision a very large phased array or parabolic antenna with up to a 200-m diameter that can produce 1000 beams, each with 2 Gbps capability. This system would represent some 8 to 10 tons in ground-based weight, including the tether structure. The projected cost of this 2 Tbps GEO platform system would be perhaps in the range of $2.5 to $3.0 billion.

This large antenna could be a more conventional parabolic shape that either flies free or is linked to power and feed systems by tethers as shown in Fig. 7.5 and identified as Concept 3-2. In light of the fact that such systems would likely not be deployed before 2012 to 2015, one might project that this large-scale antenna system would use phased-array technology and that the antenna system would be flat as depicted as shown in Fig. 7.6.

C. Concept 3-3: GEO Platform Swarm of Pico-Satellites

This is the most dramatic extension of future satellite communications performance that has been considered in the GWU-CRL study. This model essentially discards the idea of a satellite as we know it today and envisions a vast phased-array swarm of perhaps as many as 100,000 pico-satellite cells as shown in Fig. 7.7.

This swarm of pico-satellites could then flexibly create a vast number of precisely focused pencil beams that could be targeted even to specific sections of cities. These units would be naturally maintained in order by the Earth's orbital mechanics in Clarke Orbit and could be recollected by electromagnetic attraction as needed for upgrade, retrofit or end-of-life operation. This swarm would be very difficult to attack by terrorists, and electromagnetic recollection of the swarm could be accomplished on demand by activation of a magnetic switch. This swarm, as shown in Fig. 7.7 could be deployed in space over an area as large as 25 × 50 km. The feed structure would be tethered passively and be able to form a wide diversity of beams from its position as shown in Fig. 7.7.

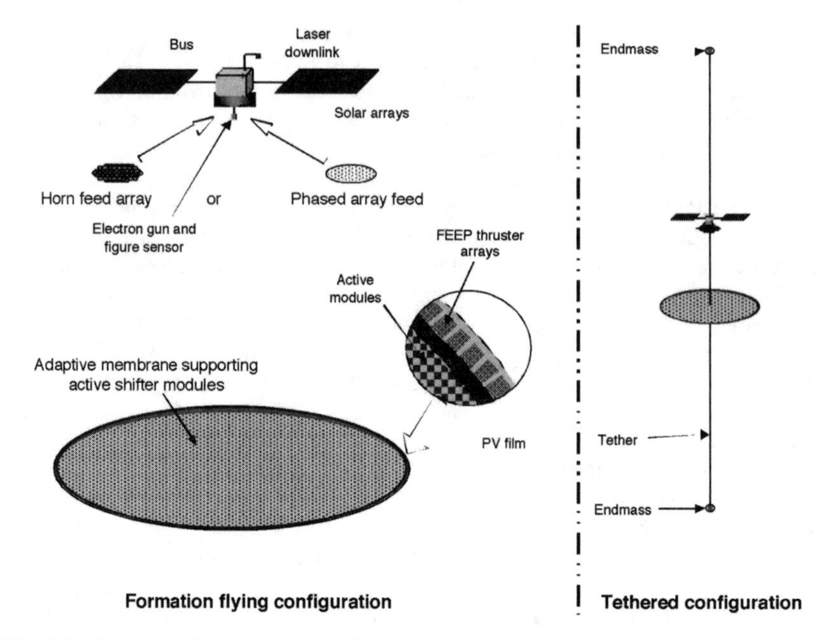

Fig. 7.6 Large-scale membrane phased-array antenna in free flying or tethered configurations (Concept 3-2-B). (Published with permission of Joseph N. Pelton and Ivan Bekey.)

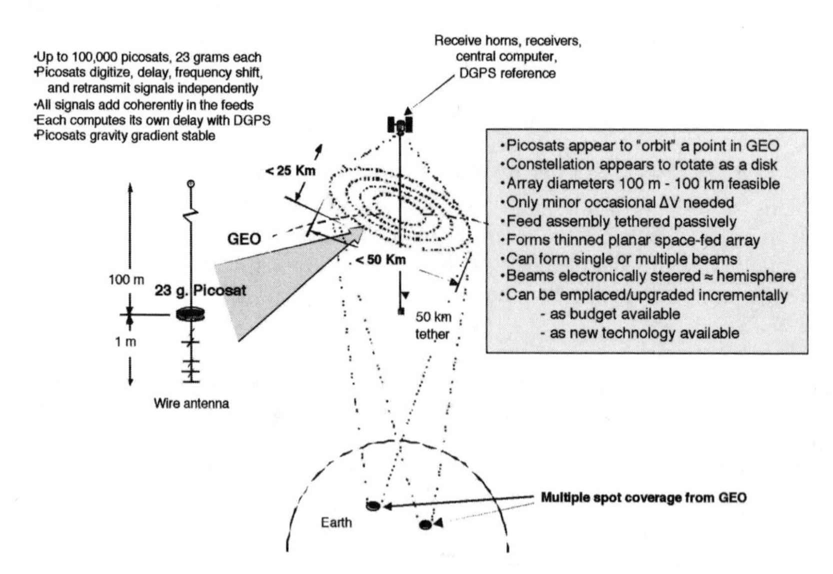

Fig. 7.7 GEO platform swarm of pico-satellite units. (Published with permission of Joseph N. Pelton and Ivan Bekey.)

The cost of this model is the most difficult to project, not only in terms of initial development cost but also in terms of the cost to design and manufacture the pico-satellite cell units and the launch and deployment costs for 100,000 or more cell units. Nevertheless, the projected cost of the launch, R&D, insurance, and production of the swarm elements would likely be on the order of $1 to $2 billion, and the performance could be as high as 4 to 5 Tbps. This concept is the most unconventional and advanced in its design. This project would involve not only a lot of future design and testing, but would also require new types of international regulations and perhaps even new types of international law to accommodate the operation of such a large-scale and distributed system.

Again, all of the technical concepts discussed in this chapter are intended only to convey new and fresh ideas about totally new designs and satellite architectures, and none of the designs as shown in Figs. 7.5–7.7 should be taken literally as complete and well-tested engineering models.

VI. Cost Estimation of the Future Communications Satellite[21]

A. Factors Defining Cost Competitiveness

Whether or not a satellite communications system is realized successfully depends on how competitive its cost is in comparison with other systems. In this section, the factors that define the cost-competitiveness of various future design concepts are discussed. For the operational services, the cost-competitiveness goal would be system designs that are less expensive than other systems that might be deployed.

1. Cost Competitiveness of Mobile Satellites

Satellite communication networks are the most appropriate ones for mobile communications because of their broad coverage and performance capabilities at a wide range of speeds, up to and including high-speed, broadband communications. One of the key factors that ultimately determines cost competitiveness is whether the service can be provided at the optimal time in terms of initial market demand. Further, the success of a mobile satellite communications system depends heavily on how early, how costly, and how "friendly" the user terminal is to the end consumer. The surprisingly rapid penetration speed of terrestrial cellular phones served to prevent the early success of the mobile satellite communications systems, especially in the case of LEO satellites. This suggests that the satellite service should be started at the optimal time. On the other hand, at this time, one or more of the GEO mobile satellite systems look as if they might be successful, at least to some degree, because even a single satellite can provide full service to an appropriately large service area.

2. Cost Competitiveness of Broadcasting Satellites

What transmission method broadcasters adopt depends in very large part on the cost of operations. In short, broadcasters generally are quite indifferent to the technology used as long as it is profitable. It is said that satellite broadcasting is very cost competitive with terrestrial broadcasting services because a single satellite can support hundreds of millions of consumers. For example, the operations cost per year per household for a satellite broadcasting system is only approximately one-fourth as much as that of a terrestrial broadcasting system, especially in areas such as Europe. Thus, a broadcasting satellite is more cost competitive than a terrestrial system. In addition, the revenues from broadcasting satellites still increase year by year as the number of subscribers is increased, as the example illustrated in Fig. 7.8 shows.[22] Therefore, the future of broadcasting satellites appears very bright. It is not likely that "Internet broadcasting" will usurp the role of broadcasting satellites, at least in the near and medium terms.

B. Cost of Future Communications Satellites

The cost of a $2G_{1S}$-satellite (whose specifications were given in Table 7.3) has been estimated. Assumptions are that the $2G_{1S}$-satellite would be a 6.8-ton-class spacecraft in weight and could embark perhaps 100 transponders. The total capacity of the satellite thus might be 60–120 Gbps and support even higher throughput rates with improved modems. The external view was shown in Fig. 7.2.

The cost of this $2G_{1S}$-satellite can be estimated as follows

1) Bus cost: $125 million.
2) Mission equipment cost: $0.58 million per transponder, including antenna. Total cost of mission equipment is $58 million.
3) Total cost: $183 million.
4) Development cost: $42 million.

The data of the three GEO platforms are listed in Table 7.5.

C. Consideration of Cost Trend

The approximate cost trend of communications satellites is shown in Fig. 7.9. This trend, however, is only based on very rough data of the satellite as estimated by early conceptual studies. The ROM cost of each satellite is shown in the upper curve, and the cost per capacity is shown by the lower curve in Fig. 7.9. These curves show that the satellite cost is approximately proportional to the $1/2$ power of the capacity. On the other hand, the cost per capacity is inversely proportional to the capacity. Thus, the cost per capacity decreases as the capacity of satellite increases. It must be particularly noted that these results are very preliminary and need to be substantiated by much further study and eventually through verification by in-orbit testing at the scale-model engineering test stage. It also needs to be emphasized that regulatory issues, health standards, and especially the development of low-cost, mobile, and highly "user friendly" terminals are critical to the successful deployment of any of the advanced GEO platform designs. There could well be serious regulatory objections to the deployment and use of any of the described designs in GEO orbit.

The cost per capacity is projected to drop to $1.5–3 million/Gbps for the $2G_{1S}$-satellite. The $2G_{1S}$-satellite has a capacity of 120 Gbps (max.). This means that 800 OC-3 (155-Mbps) links can be accommodated. If 100 subscribers with dedicated 1.5-Mbps channels are located on a 155-Mbps link, 80,000 subscribers could be accommodated; or, if this were to be on an on-demand basis, perhaps as many as a

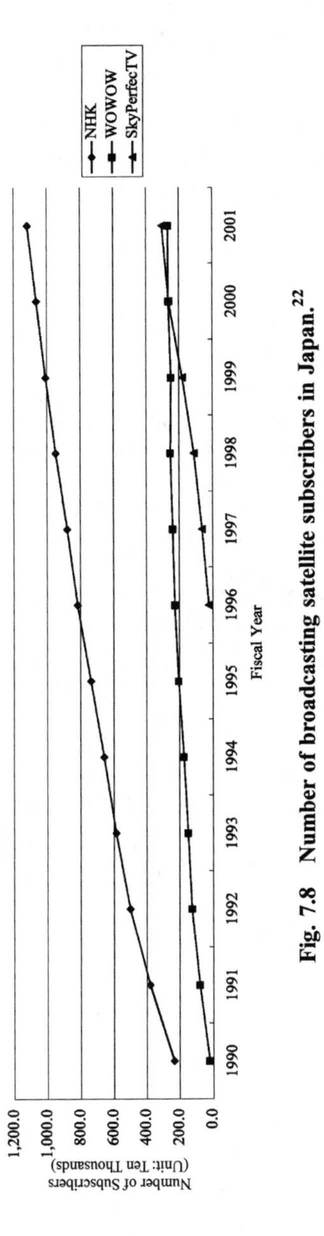

Fig. 7.8 Number of broadcasting satellite subscribers in Japan.[22]

Table 7.5 Data of GEO platforms

	Number of concept		
	Concept 3-1	Concept 3-2	Concept 3-3
Type	Tethered GEO platform network	Advanced piezo-electric shaped antennas GEO platform system	GEO platform swarm of pico-satellite units
Weight	15 ton	8–10 ton	2.3 ton
Capacity	900 Gbps	2 Tbps	4–5 Tbps
ROM cost	\$4.5 billion	\$2.5–3 billion	\$1–2 billion
Configuration	Fig. 7.4	Figs. 7.5 and 7.6	Fig. 7.7

million subscribers could be accommodated. If the investment cost of the satellite of \$183 million must be paid back over a 10-year period, then the rate of each subscriber could be as low as approximately \$20 per month for dedicated channels, or perhaps 1/10th of this for on-demand channels. This cost is only a payback of satellite cost and does not include launch cost, insurance, user terminals, development, spare satellite backup costs, or operator profits that would substantially raise the subscription rates. Nevertheless, it would seem to be sufficiently viable to deserve a detailed engineering and implementation study.

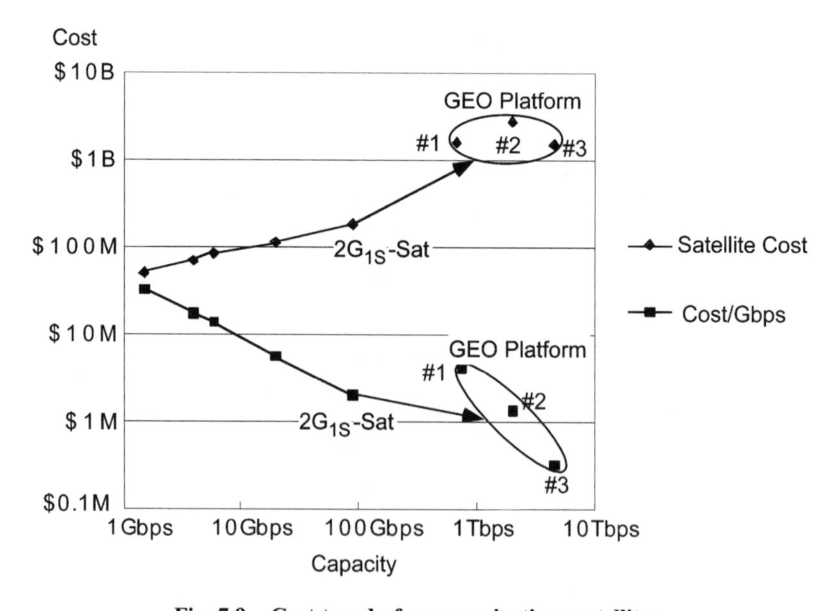

Fig. 7.9 Cost trend of communications satellites.

According to the previous examination, we can forecast that the 3G-satellite (GEO platform) would be even more cost competitive. Speaking radically, the price for each home could conceivably be much less than the 2G. The key is that programming and other value-added costs will likely determine the amount that business or consumer clients would actually pay.

D. Effective Development for the Longer Term

One of the barriers with which the satellite communications systems of the future must cope is how to provide the services at the optimal time. For the world's first type of service, it is important that the satellite communication R&D and planning be started as soon as possible. Generally speaking, it takes a long time to develop a satellite communications system, especially a satellite using new technologies. Because many new technologies need to be developed for the future large-capacity satellites, the length of development time tends to become even longer.

For the R&D satellite case, it can take 10–15 years for the most advanced technology to be developed. This lead-time depends on various factors. The major factors are enumerated in Table 7.6 using Japanese R&D satellites, but there are similar experiences in Europe, Canada, and the United States. The U.S. experiments within NASA with a "Better, Faster, Cheaper" approach to experimental satellites have had only mixed results. Most of the causes of long development times appear to be in the mission definition procedure. This procedure includes many elements, from political issues to the imputed likelihood of the successful development of the new and/or key technology program.

The development of the key technology could be accelerated by using small satellites that, because of the relatively small project cost,[23] could be launched every 1–3 years. Small satellites would be useful for a test-bed for "proof of concept," space infrastructure augmentation and test (such as formation flight), and miniaturization test of equipment. Such frequent test-bed opportunities could serve to promote the development of key technologies. Other elements that were used in the NASA Better, Faster, Cheaper program could also be examined further to seek to identify those elements that were most effective and those that adversely affected reliability and performance. The longer-term accumulation of the advanced satellite technologies via an ongoing research program, in theory, would

Table 7.6 Possible cause of long development times

Project	ETS-VI	COMETS	ETS-VIII	WINDS
Projected idea initiation	Middle of 1980s	End of 1980s	Beginning of 1990s	Beginning of 1990s
Launch year	1994	1998	2004	2005
Cause of delay				
Launcher			X	
Project budget		X		
Mission definition	X			X

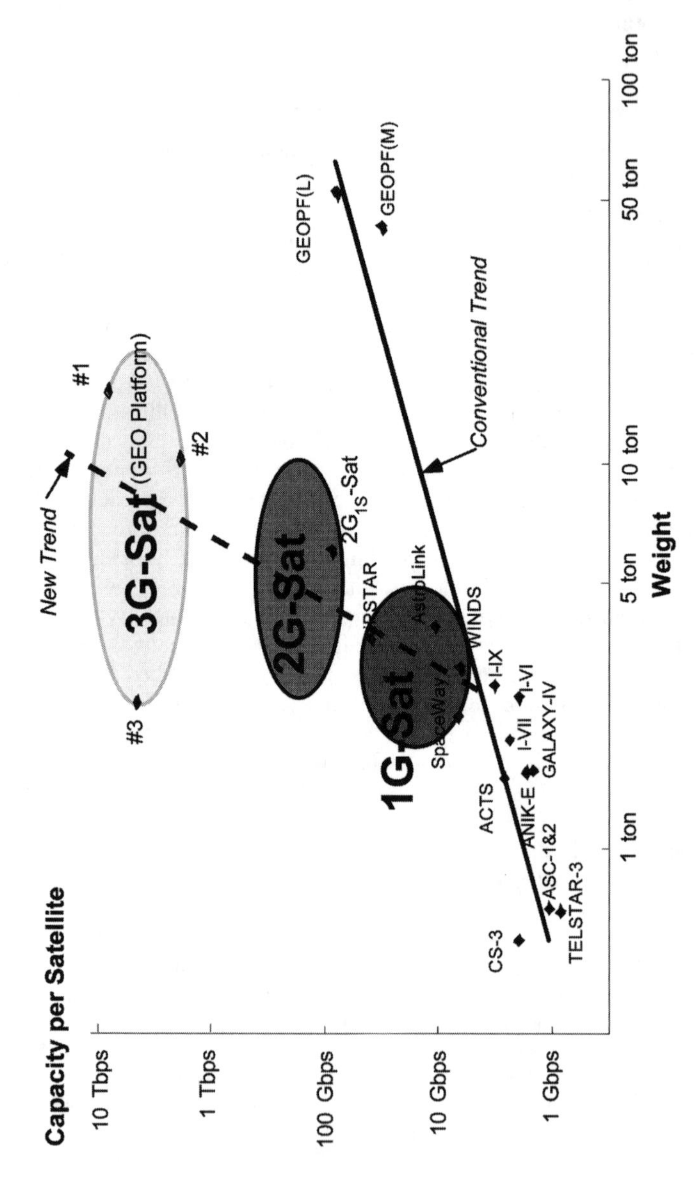

Fig. 7.10 Estimated trend of satellite weight for each generation.

help to accelerate the further development of improved, more cost effective, and more reliable communication satellites. If conducted well and in concert with the industry and academic sectors, this could also lead to the evolution of advanced and highly innovative user terminals that were responsive to the needs of business and mass-market consumers.

E. Weight of the Future Satellites

The relationship between a satellite's weight and its capacity is shown in a conventional curve in Fig. 7.10.[6] This curve includes the weight of the GEO platform that NASA studied 20 years ago. How the weight per capacity of communications satellites has decreased has heavily depended on device miniaturization. The weight of WINDS, iPSTAR, $2G_{1S}$-satellite, and three concepts of GEO platform are plotted in Fig. 7.10. As a result, the trend of satellite weight is revised, and the possibility of a new trend curve emerges.

The achievement of such a radical new trend curve depends on many factors, which include an ambitious R&D program, the realization of a supportive regulatory, technical standards, and market environment that allows these new types of satellite architectures to be developed. At this stage, it may very well be the issues of access to the GEO orbit by GEO platforms, the adoption of health standards for new types of "wearable user terminals," or other market or economic factors that will frustrate the development of this type of truly advanced system some decades from now.

VII. Impact of Expanding Human Space Activities[24]

A communication system for manned space development includes the following characteristics and needs: 1) meeting national security and legal responsibilities under UN treaties, 2) providing a fully functional global system, 3) ensuring that the system is high-speed, broadband, and extremely reliable, and 4) ensuring high-quality, deep-space performance.

As far as a manned space information/communication infrastructure is concerned, during the first generation systems, in the 2000s, the only requirements would likely be simply a global communication network. This network would need to serve as a data relay network that could support scientific and high-resolution imaging in the near-Earth environment.

In the second generation, in the 2010s, the transmission speed on the global communication network would likely need to increase substantially. In the third generation, however, in the 2020s, one might assume that manned space activity would have spread to the moon or other sustained space colonies, such as at a Lagrange point. This would imply the need for a high data rate communication network between antennas on the moon or space colony, a stationary satellite of the moon, and connection with a global satellite network.

A. Requirements for the Manned Space Communication System

Communication systems supporting manned space activities will require wideband transmission capacity, reliability, and security, including high-level encryption. Human communications requirements are obviously different from computers, and thus space communications for humans will require not only data, but voice, video, and even highly redundant audiovisual information.

Particularly in the case of long-term habitation in space, it will be necessary to provide not only numerical data but also various kinds of social communications, including connectivity to family and acquaintances, news programs, and amusement programs, to maintain normal sensitivity and psychological health of the astronauts and space-based personnel.

In the case of manned space activity, life critical messaging will be critical. The maintenance of life in a manned spacecraft, a manned space station, or even a space suit must be assured within each system. However, in the case of extraordinary events and when working far from home base, it is absolutely important to stay connected via a reliable communication link to the base and to have access to detailed information concerning all aspects of the space settlements. The reliability of the communication system for human space activity must be higher than the case of robotic and mechanical space activity, because it is a matter of life and death.

In the future, security functions will also be important in an increasingly complex number of ways. When many people are working and living in space, the protection of privacy, intellectual property, and even information about financial assets and electronic money will become an integral part of the communication system. Ensuring the security of wireless links will become very important matters.

In the age of human exploration and settlement in space, not only manned space communications will be important, but unmanned space activity supporting human activity in space will also increase in importance. For example, the use of robots controlled from the base will reduce the workload of astronauts and eliminate the need for dangerous extra-vehicular activity. In this case, the information from the robot to the operator will need to be of high quality and will need to support high-resolution imaging to accomplish sophisticated space activities. In some cases, communications to machines and robotic devices will be critical to life support systems.

Therefore, highly secure, ultra-reliable broadband communication links will likely be necessary. This implies the need to develop new technology such as virtual reality, tele-presence, or tele-existence communication systems that enable the operator to control robots or interface with machines or humans as if in their immediate presence.

B. Example of Earth–Moon Communication System

An example of communication networks for the moon is shown in Fig. 7.11. This network provides high data rate communications between data-relay satellites in a GEO orbit and moon vehicles using radio frequencies such as X-band, Ka-band, and millimeter wave, and optical frequencies. At the moon, there are three types of communication terminals—communication satellites orbiting around the moon, fixed ground stations, and mobile terminals for "autonauts" and small vehicles on the moon. In the case of a space colony, the requirements would be much simpler.

In terms of communications satellites, we cannot imagine a "lunar-stationary" satellite like a geostationary satellite in the Earth, because both gravitation forces from the Earth and the moon would interfere too greatly with the satellite orbit.

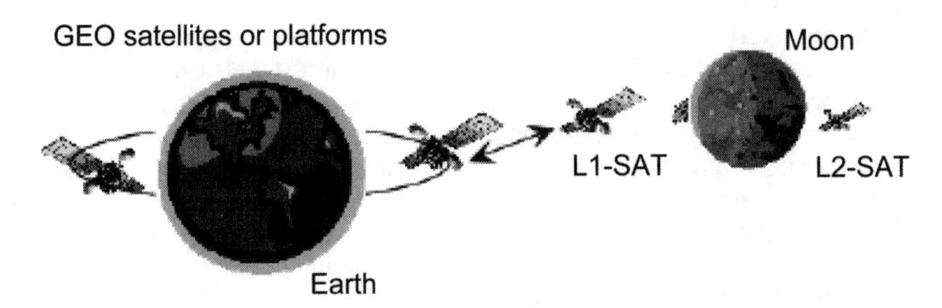

Fig. 7.11 Moon–Earth satellite communications network.

To find a stationary point for lunar satellites, we have to consider a restricted three-body problem, which gives five equilibrium points, called Lagrange points, to be found in the vicinity of the Earth and the moon. The first and second Lagrange points, L1 and L2, are located 58,390 km from the moon toward the Earth and 64,030 km from the moon on the side opposite the Earth. We will consider two satellites "parked" at the L1 and L2 points, referred to as L1-SAT and L2-SAT, (recognizing that "parked" is a misnomer because these points are unstable from the physical point of view). A ground station, referred to as a lunar base station, is located at the center of the half of the moon that faces the Earth. The vehicles running on the ground on the moon have small communications terminals or VSAT-like small terminals. Because the distances from such vehicles to L1-SAT or L2-SAT are larger than that between Earth stations and GEO satellites on the Earth, we have to make countermeasures to reduce the terminal sizes and weights. Other orbiting satellites around the moon will use L1-SAT and L2-SAT as relay satellites to communicate with the Earth terminals.

VIII. Conclusions

We have discussed the future of satellite communications, taking into account the terrestrial communications link development status and future human space activity. Satellite communications systems have, for some 20 years, progressed forward on the basis of extending and perfecting past technologies—making them more cost effective and producing higher throughput. The changing nature of market demand for broadband mobile Internet services and the challenge of fiber-optic cable technology and millimeter wave terrestrial wireless systems strongly suggest that satellite systems need to develop and deploy some new "break-through" satellite technologies in the decades ahead.

It is difficult in these rapidly changing days of accelerating technological development to envision all of the new satellite technology that will unfold during the next 30 years. Certainly we can envision that there will be an integration of all forms of digital communications with other information functions, such as space navigation and position determination, and computational and text-related functions. Indeed, the integration of all these functions into user terminal devices that are increasingly miniaturized and "user friendly" may be the most important force in future satellite communications market-related trends.

Thus, the concepts of future satellite communications development provided in this chapter should be considered in only conceptual terms that seek only to point toward appropriate future capabilities. Even so, it is important for us to identify futuristic and challenging targets. One can plan for the future by extrapolating from past trends or create new normative goals to create new futures. It is hoped that some of the technological scenarios set forth in this chapter can help inspire a new and more dynamic future for the satellite industry several decades into the future. The concepts discussed here are only an attempt to develop some totally new models for 21st century satellite networking.

References

[1]Bates, J., "Satellite Internet Market Expanding Rapidly," *Space News*, Vol. 11, No. 16, p. 4.

[2]Taverna, M. A., "Industry Bullish on Broadband, But Impact on Satellites in Doubt," *Aviation Week & Space Technology*, Vol. 153, No. 16, 16 Oct. 2000, pp. 55–57; also Pelton, J. N., *The New Satellite Industry: Revenue Opportunities and Opportunities for Success*, Chicago, International Engineering Consortium, 2002.

[3]Helm, N. R., "*Satellite Communications for the Internet*," *Proceedings of the 18th AIAA International Communications Satellite Systems Conference*, AIAA, Reston, VA, 2000; also AIAA Paper 2000-1165.

[4]Foley, T., "Satellite and ISPs Carving Out a Niche Service," *Via Satellite*, Vol. XV, No. 11, Nov. 2000, pp. 18–26.

[5]Silverstein, S., "Internet Prompts Satellite Firms to Re-evaluate Plans," *Space News*, Vol. 11, No. 48, 18 Dec. 2000, p. 4.

[6]Iida, T., and Pelton, J. N., "Conceptual Study of Advanced Asia-Pacific Telecommunications Satellite for Future Gigabit Transmission," *Space Communications*, Vol. 11, No. 3, 1993, pp. 193–203.

[7]Hsu, E., Wang, C., Bargman, L., Kadowaki, N., Yoshimura, N., Takahashi, T., Pearman, J., Gargione, F., Bhasin, K., Gary, O., Vlark, G., Shopbell, P., Yoshikawa, M., Gill, M., Tatsumi, H., and desJardins, R., "Global Interoperability of High Definition Video Streams Via ACTS and INTELSAT," *Proceedings of the 6th Ka Band Utilization Conference*, 31 May–2 June, 2000, pp. 123–130.

[8]Gedney, R. T., Schertler, R., and Gargione, F., *The Advanced Communications Technology Satellite*, Scitech Publishing, Inc., 2000.

[9]Edelson, B. I., "Commentary," *Space News*, 11 Sept. 2000.

[10]Iida, T., and Suzuki, Y., "International Experimentation Aspect of Satellite Communications Toward GII," *50th International Astronautical Congress*, Amsterdam, The Netherlands, IAF-99-M.1.05, 4 Oct. 1999; also, *Space Technology*, Lister Science, Vol. 20, No. 5–6, 2001, pp. 187–198.

[11]*Okinawa Charter on Global Information Society*, Okinawa G-8 Summit 2000, June 2000.

[12]Iida, T., and Suzuki, Y., "Satellite Communications R&D for Next 30 Years," *AIAA 19th International Communications Satellite Systems Conference*, Toulouse, France, No. 233, April 2001; also, *Space Communications*, No. 3, Vol. 17, No. 4, 2001, pp. 271–277; Pelton, J. N., *The New Satellite Industry: Revenue Opportunities and Opportunities for Success*, Chicago, International Engineering Consortium, 2002.

[13]Kuramasu, R., Araki, T., Shimada, M., Tomita, E., Satoh, T., Kuroda, T., Yajima, M., Maeda, T., Mukai, T., Kadowaki, N., and Nakao, M., "The Wideband Internetworking

Engineering Test And Demonstration Satellite (WINDS) System," *AIAA 20th International Communications Satellite Systems Conference*, AIAA, Reston, VA, May 2002; also AIAA Paper 2002–2044.
[14]Suzuki, R., Nishiyama, I., Motoyoshi, S., Morikawa, E., and Yasuda, Y., "Current Status of Nels Project: R&D of Global Multimedia Mobile Satellite Communications," *AIAA 20th International Communications Satellite Systems Conference*, AIAA, Reston, VA, May 2002; also AIAA Paper 2002–2038.
[15]Iida, T., Arimoto, Y., Suzuki, Y., and Bhasin, K. B., "A Consideration on the Inter-Satellite Link for a Gigabit Information Space Corridor," *AIAA 17th International Communications Satellite Systems Conference and Exhibit*, AIAA, Reston, VA, Feb 1998; also AIAA Paper 98–1241.
[16]MPT Telecommunication Technology Council, "Basic Plan of Information Communication R&D Policy for R&D of Information Communication Technology in the 21st Century," Feb. 2000.
[17]"Introduction to Shin Satellete PLC," *Shin Satellite*, 2002.
[18]Iida, T., Suzuki, Y., and Akaishi, A., "Communications Satellite R&D for Next 30 Years: Follow-On," *AIAA 20th International Communications Satellite Systems Conference*, AIAA, Reston, VA, May 2002; also AIAA Paper 2002-1972.
[19]Pelton, J. N., and Bekey, I., "Advanced Geoplatform Concepts," *TSMMW 2002 Conference*, 4-4, Yokusuka, Japan, 14–15 March 2002.
[20]Pelton, J. N., Bekey, I., and Edelson, B. E., "Phase 3 Report on Advanced Geoplatform Satellite Concepts," George Washington Univ., Jan. 2002.
[21]Iida, T., "Cost Consideration for Future Communications Satellite," *53rd International Astronautical Congress, The World Space Congress 2002*, IAC-02-IAA.1.4.05, 10–19 Oct. 2002.
[22]"Info-Communications White Paper of FY 2002," Ministry of Public Management, Home Affairs, Posts and Telecommunications, 2002.
[23]Iida, T., "Creation of Information Society by Satellite Communications R&D, Role of Small Satellites," *6th ISU Annual International Symposium, Smaller Satellites, Bigger Business?*, 21 May 2001; also, Rycroft, M., and Crosby, N. (eds.), *Smaller Satellites, Bigger Business? Concepts, Applications and Markets for Micro/Nanosatellites in a New Information World*, Kluwer Academic Publishers, 2002, pp. 93–100.
[24]Suzuki, Y., Wakana, H., and Iida, T., "Future Vision of Satellite Communications for Expanding Human Activities," *52nd International Astronautical Congress*, IAF-01-M.1.0.2, Oct. 2001.
[25]Iida, T. (ed.), "Satellite Communications System and Its Design Technology," Ohmsha, Ltd. and IOS Press, Tokyo, Japan, 2000.

Questions for Discussion

1) **The orbital altitude of 15,500 km was selected in the SISCEO shown in Fig. 7.3. Why was such an orbit selected and how might this be further optimized?**

To examine a point-to-point satellite communications link, Tokyo was assumed as an example of the typical Earth station location in the Northern Hemisphere and a 30-deg elevation angle was selected as a minimal angle for acceptable broadband service. Tokyo is located at 36° North latitude.

Figure 7.12 shows the number of satellites on the equatorial orbit vs. orbit altitude to maintain the minimum elevation angle of 30 deg. Figure 7.12 shows that having more than six satellites does not decrease the orbital altitude dramatically, and less than six requires increasing the altitude very quickly. In this case, the variation of slant range between Earth station and satellite is 17,974–17,108 km. The delay time for a round-trip propagation is approximately 120 ms, which is about half that of a GEO satellite system. Thus, the 15,500 km altitude orbit was chosen. Consider the case wherein a somewhat higher orbit and fewer satellites might be used to carry out a sensitivity or optimization study. If one were to use more satellites in lower orbit would the van Allen Belts become a consideration in your calculations? And if so at what altitude? Finally consider the elevation angles for this service for Santiago, Chile to Stockholm, Sweden.

2) **One of the key elements of satellite communications is the issue of transmission delay. The idea of an optical ring of satellites is to reduce latency in broadband space communications systems. Consider methods as to how latency might be further reduced starting with the following considerations taken from the optical ring at 15,500 km altitude.**

One can start by taking Tokyo and Los Angeles as references location. We can show the effectiveness of using an orbit of 15,500 km altitude as

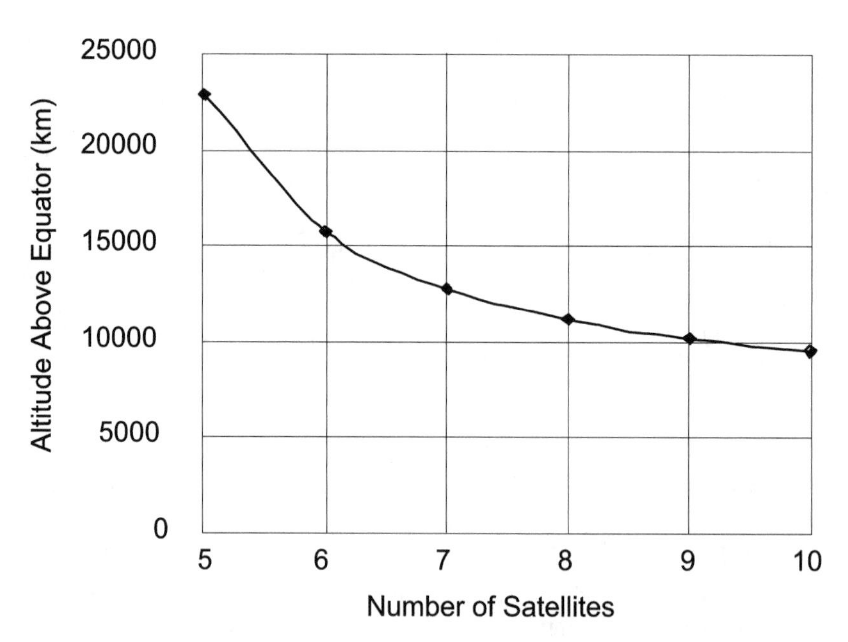

Fig. 7.12 Number of satellites vs. orbit height, when minimum elevation angle is 30 deg.

follows. We consider the case of a satellite communications link between Tokyo and Los Angeles. If we use two non-GEO satellites of 15,500 km altitude, separated 60° in longitude and linked up via ISL, the distance from Tokyo to Los Angeles is 57,478 km 192 ms), the value in the parenthesis means propagation delay time, while one GEO 78,878 km (263 ms). The propagation delay is less by 71 ms, which means a 27% decrease in the propagation delay. The terrestrial distance between them is approximately 9772 km (33 ms). The actual propagation delay over the terrestrial line is absolutely much larger than 33 ms, probably more than 100 ms estimating. Furthermore, if we consider the communication link between Tokyo and any city located more interior to the North American continent than Los Angeles, we can estimate that the delay in the terrestrial-fiber network may be dramatically increased because of having to pass through many switching circuits. Therefore, the non-GEO satellite system having an equatorial orbit of 15,500 km altitude could be competitive with the terrestrial optical-fiber network even in propagation delay.

3) **Concerning the communications link between the moon and Earth, we consider GEO satellites, or GEO platforms orbiting around a GEO orbit for global and continuous connections in the Earth side. Calculate an RF link directly between the moon's visible side to Earth. Then consider the case of RF links from the moon to the L-1 or L-2 satellite and from these satellites to the Earth.**

These satellites are assumed to have 3.6-m-diam antenna and a G/T of 28.8 dB/K. Ka-band 32 GHz) is assumed for link calculations. In the moon-to-Earth link, when both L1-SAT and L2-SAT have 3.6-m-diam antennas with a transmitting power of 30 W, link margins[25] for 156 Mbps transmissions are 6.9 dB and 4.1 dB for L1-SAT and L2-SAT, respectively. Moreover, when we use lunar base stations with an antenna diameter of 10 m and 3.6 m, we will get larger margins such as 14.3 dB and 5.4 dB, respectively. On the other hand, in the Earth-to-moon link, it is assumed that a GEO satellite has a 3.6-m-diam antenna with a transmitting power of 30 W, and both L1-SAT and L2-SAT have 3.6-m-diam antennas with a G/T of 28.8 dB/K at 32 GHz. The link calculation gives link margins of 6.9 dB for L1-SAT link and of 4.1 dB for L2-SAT link at 156 Mbps. When the lunar station is used, link margins of 14.3 dB for 10-m antennas and of 5.4 dB for 3.6-m antennas will be available. In both Earth-to-moon and moon-to-Earth links, satellite communications terminals with antennas larger than 3.6 m in diameter can achieve high-data transmissions of 156 Mbps. The L2-SAT cannot see the Earth itself because of blockage by the moon, but can see GEO satellites. Several GEO satellites will be needed for continuous connections between L2-SAT and the Earth. Small vehicle terminals with an antenna of 75 cm in diameter and a transmitting power of 10 W for 156 Mbps communications will be available with margins of 3.1 dB for L1-SAT and 2.2 dB for L2-SAT.

4) **The era of very inexpensive subscriber fees as mentioned above is highly dependent on market demand and the evolution of satellite specific applications. Discuss what new types of service will emerge in the future, for example, by 2010 and then by 2020?**

   The evolution of telecommunications services has remained exponential for more than 100 years. The dramatic surge in new high data rate capabilities in the past decades seems to have been stimulated by breakthroughs in fiber-optic technology and the development of dense wave division multiplexing. This technology, however, is not well suited to broadband mobile services. What kinds of new mobile services can be anticipated? Why is it that satellite multiplexing systems are focused on systems that work in the time domain such as CDMA and TDMA) while fiber-optic systems are focused on frequency or wave division systems? Will this be the source of a continuing compatibility problem between satellites and fiber networks?

5) **Discuss what satellite transmission method is feasible to establish over 100 Gbps throughput and what problems would need to be solved to implement this as an operational service. Further, this might give rise to special concerns about satellite latency and specific protocol-related issues such as those related to IP Sec.**

   In addressing this issue, consider such factors as new satellite modem technology, the size and characteristics of the satellite antennas, the size and characteristics of the spot beams that might be used in terms of catchment areas on the ground, the spectrum band that would be used for example, C, Ku, Ka, or Q/V), and the size and characteristics of the user terminals. When do you think that this type of satellite service might be introduced and what is the cost of service that might apply to provide broadband services such as at 155.5 Mbps? What can be done to address such issues as satellite latency or special requirements related IP protocol issues and VPN applications?

6) **Three types of designs for advanced GEO platform satellites were described in broad terms in this chapter. Address the pros and cons of each design in terms of satellite technology, ground terminals, regulatory or standards issues, and overall feasibility.**

   The discussion of satellite technology should consider new materials especially piezo-electric membranes), tether technology, stabilization techniques, solar cell and electrical power considerations, TWTA vs. SSPA, and the feasibility of in-orbit testing using scale model engineering prototypes. To what extent do you see the demand for new services and the development of new types of "personal" broadband user terminals driving the development of advanced GEO platform systems? Can you foresee legal, regulatory, political, health standards, or other issues blocking the deployment of such large-scale systems? Seek to rate all of these design concepts based on at least ten different criteria.

Index

PROGRESS IN ASTRONAUTICS AND AERONAUTICS SERIES VOLUMES

***1. Solid Propellant Rocket Research (1960)**
Martin Summerfield
Princeton University

***2. Liquid Rockets and Propellants (1960)**
Loren E. Bollinger
Ohio State University
Martin Goldsmith
The Rand Corp.
Alexis W. Lemmon Jr.
Battelle Memorial Institute

***3. Energy Conversion for Space Power (1961)**
Nathan W. Snyder
Institute for Defense Analyses

***4. Space Power Systems (1961)**
Nathan W. Snyder
Institute for Defense Analyses

***5. Electrostatic Propulsion (1961)**
David B. Langmuir
Space Technology Laboratories, Inc.
Ernst Stuhlinger
NASA George C. Marshall Space Flight Center
J. M. Sellen Jr.
Space Technology Laboratories, Inc.

***6. Detonation and Two-Phase Flow (1962)**
S. S. Penner
California Institute of Technology
F. A. Williams
Harvard University

***7. Hypersonic Flow Research (1962)**
Frederick R. Riddell
AVCO Corp.

***8. Guidance and Control (1962)**
Robert E. Roberson
Consultant
James S. Farrior
Lockheed Missiles and Space Co.

***9. Electric Propulsion Development (1963)**
Ernst Stuhlinger
NASA George C. Marshall Space Flight Center

***10. Technology of Lunar Exploration (1963)**
Clifford I. Cumming
Harold R. Lawrence
Jet Propulsion Laboratory

***11. Power Systems for Space Flight (1963)**
Morris A. Zipkin
Russell N. Edwards
General Electric Co.

***12. Ionization in High-Temperature Gases (1963)**
Kurt E. Shuler, Editor
National Bureau of Standards
John B. Fenn,
Associate Editor
Princeton University

***13. Guidance and Control–II (1964)**
Robert C. Langford
General Precision Inc.
Charles J. Mundo
Institute of Naval Studies

***14. Celestial Mechanics and Astrodynamics (1964)**
Victor G. Szebehely
Yale University Observatory

***15. Heterogeneous Combustion (1964)**
Hans G. Wolfhard
Institute for Defense Analyses
Irvin Glassman
Princeton University
Leon Green Jr.
Air Force Systems Command

***16. Space Power Systems Engineering (1966)**
George C. Szego
Institute for Defense Analyses
J. Edward Taylor
TRW Inc.

***17. Methods in Astrodynamics and Celestial Mechanics (1966)**
Raynor L. Duncombe
U.S. Naval Observatory
Victor G. Szebehely
Yale University Observatory

***18. Thermophysics and Temperature Control of Spacecraft and Entry Vehicles (1966)**
Gerhard B. Heller
NASA George C. Marshall Space Flight Center

***19. Communication Satellite Systems Technology (1966)**
Richard B. Marsten
Radio Corporation of America

---
*Out of print.

***20. Thermophysics of Spacecraft and Planetary Bodies: Radiation Properties of Solids and the Electromagnetic Radiation Environment in Space (1967)**
Gerhard B. Heller
NASA George C. Marshall Space Flight Center

***21. Thermal Design Principles of Spacecraft and Entry Bodies (1969)**
Jerry T. Bevans
TRW Systems

***22. Stratospheric Circulation (1969)**
Willis L. Webb
Atmospheric Sciences Laboratory, White Sands, and University of Texas at El Paso

***23. Thermophysics: Applications to Thermal Design of Spacecraft (1970)**
Jerry T. Bevans
TRW Systems

***24. Heat Transfer and Spacecraft Thermal Control (1971)**
John W. Lucas
Jet Propulsion Laboratory

***25. Communication Satellites for the 70's: Technology (1971)**
Nathaniel E. Feldman
The Rand Corp.
Charles M. Kelly
The Aerospace Corp.

***26. Communication Satellites for the 70's: Systems (1971)**
Nathaniel E. Feldman
The Rand Corp.
Charles M. Kelly
The Aerospace Corp.

***27. Thermospheric Circulation (1972)**
Willis L. Webb
Atmospheric Sciences Laboratory, White Sands, and University of Texas at El Paso

***28. Thermal Characteristics of the Moon (1972)**
John W. Lucas
Jet Propulsion Laboratory

***29. Fundamentals of Spacecraft Thermal Design (1972)**
John W. Lucas
Jet Propulsion Laboratory

***30. Solar Activity Observations and Predictions (1972)**
Patrick S. McIntosh
Murray Dryer
Environmental Research Laboratories, National Oceanic and Atmospheric Administration

***31. Thermal Control and Radiation (1973)**
Chang-Lin Tien
University of California at Berkeley

***32. Communications Satellite Systems (1974)**
P. L. Bargellini
COMSAT Laboratories

***33. Communications Satellite Technology (1974)**
P. L. Bargellini
COMSAT Laboratories

***34. Instrumentation for Airbreathing Propulsion (1974)**
Allen E. Fuhs
Naval Postgraduate School
Marshall Kingery
Arnold Engineering Development Center

***35. Thermophysics and Spacecraft Thermal Control (1974)**
Robert G. Hering
University of Iowa

***36. Thermal Pollution Analysis (1975)**
Joseph A. Schetz
Virginia Polytechnic Institute
ISBN 0-915928-00-0

***37. Aeroacoustics: Jet and Combustion Noise; Duct Acoustics (1975)**
Henry T. Nagamatsu, Editor
General Electric Research and Development Center
Jack V. O'Keefe, Associate Editor
The Boeing Co.
Ira R. Schwartz, Associate Editor
NASA Ames Research Center
ISBN 0-915928-01-9

***38. Aeroacoustics: Fan, STOL, and Boundary Layer Noise; Sonic Boom; Aeroacoustics Instrumentation (1975)**
Henry T. Nagamatsu, Editor
General Electric Research and Development Center
Jack V. O'Keefe, Associate Editor
The Boeing Co.
Ira R. Schwartz, Associate Editor
NASA Ames Research Center
ISBN 0-915928-02-7

***39. Heat Transfer with Thermal Control Applications (1975)**
M. Michael Yovanovich
University of Waterloo
ISBN 0-915928-03-5

*Out of print.

***40. Aerodynamics of Base Combustion (1976)**
S. N. B. Murthy, Editor
J. R. Osborn, Associate Editor
Purdue University
A. W. Barrows
J. R. Ward,
Associate Editors
Ballistics Research Laboratories
ISBN 0-915928-04-3

***41. Communications Satellite Developments: Systems (1976)**
Gilbert E. LaVean
Defense Communications Agency
William G. Schmidt
CML Satellite Corp.
ISBN 0-915928-05-1

***42. Communications Satellite Developments: Technology (1976)**
William G. Schmidt
CML Satellite Corp.
Gilbert E. LaVean
Defense Communications Agency
ISBN 0-915928-06-X

***43. Aeroacoustics: Jet Noise, Combustion and Core Engine Noise (1976)**
Ira R. Schwartz, Editor
NASA Ames Research Center
Henry T. Nagamatsu,
Associate Editor
General Electric Research and Development Center
Warren C. Strahle,
Associate Editor
Georgia Institute of Technology
ISBN 0-915928-07-8

***44. Aeroacoustics: Fan Noise and Control; Duct Acoustics; Rotor Noise (1976)**
Ira R. Schwartz, Editor
NASA Ames Research Center
Henry T. Nagamatsu,
Associate Editor
General Electric Research and Development Center
Warren C. Strahle,
Associate Editor
Georgia Institute of Technology
ISBN 0-915928-08-6

***45. Aeroacoustics: STOL Noise; Airframe and Airfoil Noise (1976)**
Ira R. Schwartz, Editor
NASA Ames Research Center
Henry T. Nagamatsu,
Associate Editor
General Electric Research and Development Center
Warren C. Strahle,
Associate Editor
Georgia Institute of Technology
ISBN 0-915928-09-4

***46. Aeroacoustics: Acoustic Wave Propagation; Aircraft Noise Prediction; Aeroacoustic Instrumentation (1976)**
Ira R. Schwartz, Editor
NASA Ames Research Center
Henry T. Nagamatsu,
Associate Editor
General Electric Research and Development Center
Warren C. Strahle,
Associate Editor
Georgia Institute of Technology
ISBN 0-915928-10-8

***47. Spacecraft Charging by Magnetospheric Plasmas (1976)**
Alan Rosen
TRW Inc.
ISBN 0-915928-11-6

***48. Scientific Investigations on the Skylab Satellite (1976)**
Marion I. Kent
Ernst Stuhlinger
NASA George C. Marshall Space Flight Center
Shi-Tsan Wu
University of Alabama
ISBN 0-915928-12-4

***49. Radiative Transfer and Thermal Control (1976)**
Allie M. Smith
ARO Inc.
ISBN 0-915928-13-2

***50. Exploration of the Outer Solar System (1976)**
Eugene W. Greenstadt
TRW Inc.
Murray Dryer
National Oceanic and Atmospheric Administration
Devrie S. Intriligator
University of Southern California
ISBN 0-915928-14-0

***51. Rarefied Gas Dynamics, Parts I and II (two volumes) (1977)**
J. Leith Potter
ARO Inc.
ISBN 0-915928-15-9

***52. Materials Sciences in Space with Application to Space Processing (1977)**
Leo Steg
General Electric Co.
ISBN 0-915928-16-7

---
*Out of print.

***53. Experimental Diagnostics in Gas Phase Combustion Systems (1977)**
Ben T. Zinn, Editor
Georgia Institute of Technology
Craig T. Bowman, Associate Editor
Stanford University
Daniel L. Hartley, Associate Editor
Sandia Laboratories
Edward W. Price, Associate Editor
Georgia Institute of Technology
James G. Skifstad, Associate Editor
Purdue University
ISBN 0-915928-18-3

***54. Satellite Communication: Future Systems (1977)**
David Jarett
TRW Inc.
ISBN 0-915928-18-3

***55. Satellite Communications: Advanced Technologies (1977)**
David Jarett
TRW Inc.
ISBN 0-915928-19-1

***56. Thermophysics of Spacecraft and Outer Planet Entry Probes (1977)**
Allie M. Smith
ARO Inc.
ISBN 0-915928-20-5

***57. Space-Based Manufacturing from Nonterrestrial Materials (1977)**
Gerald K. O'Neill, Editor
Brian O'Leary, Assistant Editor
Princeton University
ISBN 0-915928-21-3

***58. Turbulent Combustion (1978)**
Lawrence A. Kennedy
State University of New York at Buffalo
ISBN 0-915928-22-1

***59. Aerodynamic Heating and Thermal Protection Systems (1978)**
Leroy S. Fletcher
University of Virginia
ISBN 0-915928-23-X

***60. Heat Transfer and Thermal Control Systems (1978)**
Leroy S. Fletcher
University of Virginia
ISBN 0-915928-24-8

***61. Radiation Energy Conversion in Space (1978)**
Kenneth W. Billman
NASA Ames Research Center
ISBN 0-915928-26-4

***62. Alternative Hydrocarbon Fuels: Combustion and Chemical Kinetics (1978)**
Craig T. Bowman
Stanford University
Jorgen Birkeland
Department of Energy
ISBN 0-915928-25-6

***63. Experimental Diagnostics in Combustion of Solids (1978)**
Thomas L. Boggs
Naval Weapons Center
Ben T. Zinn
Georgia Institute of Technology
ISBN 0-915928-28-0

***64. Outer Planet Entry Heating and Thermal Protection (1979)**
Raymond Viskanta
Purdue University
ISBN 0-915928-29-9

***65. Thermophysics and Thermal Control (1979)**
Raymond Viskanta
Purdue University
ISBN 0-915928-30-2

***66. Interior Ballistics of Guns (1979)**
Herman Krier
University of Illinois at UrbanaChampaign
Martin Summerfield
New York University
ISBN 0-915928-32-9

***67. Remote Sensing of Earth from Space: Role of Smart Sensors (1979)**
Roger A. Breckenridge
NASA Langley Research Center
ISBN 0-915928-33-7

***68. Injection and Mixing in Turbulent Flow (1980)**
Joseph A. Schetz
Virginia Polytechnic Institute and State University
ISBN 0-915928-35-3

*Out of print.

***69. Entry Heating and Thermal Protection (1980)**
Walter B. Olstad
NASA Headquarters
ISBN 0-915928-38-8

***70. Heat Transfer, Thermal Control, and Heat Pipes (1980)**
Walter B. Olstad
NASA Headquarters
ISBN 0-915928-39-6

***71. Space Systems and Their Interactions with Earth's Space Environment (1980)**
Henry B. Garrett
Charles P. Pike
Hanscom Air Force Base
ISBN 0-915928-41-8

***72. Viscous Flow Drag Reduction (1980)**
Gary R. Hough
Vought Advanced Technology Center
ISBN 0-915928-44-2

***73. Combustion Experiments in a Zero-Gravity Laboratory (1981)**
Thomas H. Cochran
NASA Lewis Research Center
ISBN 0-915928-48-5

***74. Rarefied Gas Dynamics, Parts I and II (two volumes) (1981)**
Sam S. Fisher
University of Virginia
ISBN 0-915928-51-5

***75. Gasdynamics of Detonations and Explosions (1981)**
J. R. Bowen
University of Wisconsin at Madison
N. Manson
Universite de Poitiers
A. K. Oppenheim
University of California at Berkeley
R. I. Soloukhin
Institute of Heat and Mass Transfer, BSSR Academy of Sciences
ISBN 0-915928-46-9

***76. Combustion in Reactive Systems (1981)**
J. R. Bowen
University of Wisconsin at Madison
N. Manson
Universite de Poitiers
A. K. Oppenheim
University of California at Berkeley
R. I. Soloukhin
Institute of Heat and Mass Transfer, BSSR Academy of Sciences
ISBN 0-915928-47-7

***77. Aerothermodynamics and Planetary Entry (1981)**
A. L. Crosbie
University of Missouri-Rolla
ISBN 0-915928-52-3

***78. Heat Transfer and Thermal Control (1981)**
A. L. Crosbie
University of Missouri-Rolla
ISBN 0-915928-53-1

***79. Electric Propulsion and Its Applications to Space Missions (1981)**
Robert C. Finke
NASA Lewis Research Center
ISBN 0-915928-55-8

***80. Aero-Optical Phenomena (1982)**
Keith G. Gilbert
Leonard J. Otten
Air Force Weapons Laboratory
ISBN 0-915928-60-4

***81. Transonic Aerodynamics (1982)**
David Nixon
Nielsen Engineering & Research, Inc.
ISBN 0-915928-65-5

***82. Thermophysics of Atmospheric Entry (1982)**
T. E. Horton
University of Mississippi
ISBN 0-915928-66-3

***83. Spacecraft Radiative Transfer and Temperature Control (1982)**
T. E. Horton
University of Mississippi
ISBN 0-915928-67-1

***84. Liquid-Metal Flows and Magneto-hydrodynamics (1983)**
H. Branover
Ben-Gurion University of the Negev
P. S. Lykoudis
Purdue University
A. Yakhot
Ben-Gurion Universityof the Negev
ISBN 0-915928-70-1

*Out of print.

***85. Entry Vehicle Heating and Thermal Protection Systems: Space Shuttle, Solar Starprobe, Jupiter Galileo Probe (1983)**
Paul E. Bauer
McDonnell Douglas Astronautics Co.
Howard E. Collicott
The Boeing Co.
ISBN 0-915928-74-4

***86. Spacecraft Thermal Control, Design, and Operation (1983)**
Howard E. Collicott
The Boeing Co.
Paul E. Bauer
McDonnell Douglas Astronautics Co.
ISBN 0-915928-75-2

***87. Shock Waves, Explosions, and Detonations (1983)**
J. R. Bowen
University of Washington
N. Manson
Universite de Poitiers
A. K. Oppenheim
University of California at Berkeley
R. I. Soloukhin
Institute of Heat and Mass Transfer, BSSR Academy of Sciences
ISBN 0-915928-76-0

***88. Flames, Lasers, and Reactive Systems (1983)**
J. R. Bowen
University of Washington
N. Manson
Universite de Poitiers
A. K. Oppenheim
University of California at Berkeley
R. I. Soloukhin
Institute of Heat and Mass Transfer, BSSR Academy of Sciences
ISBN 0-915928-77-9

***89. Orbit-Raising and Maneuvering Propulsion: Research Status and Needs (1984)**
Leonard H. Caveny
Air Force Office of Scientific Research
ISBN 0-915928-82-5

***90. Fundamentals of Solid-Propellant Combustion (1984)**
Kenneth K. Kuo
Pennsylvania State University
Martin Summerfield
Princeton Combustion Research Laboratories, Inc.
ISBN 0-915928-84-1

***91. Spacecraft Contamination: Sources and Prevention (1984)**
J. A. Roux, Editor
University of Mississippi
T. D. McCay, Editor
NASA Marshall Space Flight Center
ISBN 0-915928-85-X

***92. Combustion Diagnostics by Nonintrusive Methods (1984)**
T. D. McCay, Editor
NASA Marshall Space Flight Center
J. A. Roux, Editor
University of Mississippi
ISBN 0-915928-86-8

***93. The INTELSAT Global Satellite System (1984)**
Joel Alper
COMSAT Corp.
Joseph Pelton
INTELSAT
ISBN 0-915928-90-6

***94. Dynamics of Shock Waves, Explosions, and Detonations (1984)**
J. R. Bowen
University of Washington
N. Manson
Universite de Poitiers
A. K. Oppenheim
University of California at Berkeley
R. I. Soloukhin
Institute of Heat and Mass Transfer, BSSR Academy of Sciences
ISBN 0-915928-91-4

***95. Dynamics of Flames and Reactive Systems (1984)**
J. R. Bowen
University of Washington
N. Manson
Universite de Poitiers
A. K. Oppenheim
University of California at Berkeley
R. I. Soloukhin
Institute of Heat and Mass Transfer, BSSR Academy of Sciences
ISBN 0-915928-92-2

***96. Thermal Design of Aeroassisted Orbital Transfer Vehicles (1985)**
H. F. Nelson, Editor
University of Missouri-Rolla
ISBN 0-915928-94-9

***97. Monitoring Earth's Ocean, Land, and Atmosphere from Space—Sensors, Systems, and Applications (1985)**
Abraham Schnapf
Aerospace Systems Engineering
ISBN 0-915928-98-1

*Out of print.

98. Thrust and Drag: Its Prediction and Verification (1985)
Eugene E. Covert
Massachusetts Institute of Technology
C. R. James
Vought Corp.
William F. Kimzey
Sverdrup Technology AEDC Group
George K. Richey
U.S. Air Force
Eugene C. Rooney
U.S. Navy Department of Defense
ISBN 0-930403-00-2

99. Space Stations and Space Platforms—Concepts, Design, Infrastructure, and Uses (1985)
Ivan Bekey
Daniel Herman
NASA Headquarters
ISBN 0-930403-01-0

***100. Single- and Multi-Phase Flows in an Electromagnetic Field: Energy, Metallurgical, and Solar Applications (1985)**
Herman Branover, Editor
Ben-Gurion University of the Negev
Paul S. Lykoudis, Editor
Purdue University
Michael Mond, Editor
Ben-Gurion University of the Negev
ISBN 0-930403-04-5

***101. MHD Energy Conversion: Physiotechnical Problems (1986)**
V. A. Kirillin, Editor
A. E. Sheyndlin, Editor
Soviet Academy of Sciences
ISBN 0-930403-05-3

***102. Numerical Methods for Engine-Airframe Integration (1986)**
S. N. B. Murthy, Editor
Purdue University
Gerald C. Paynter, Editor
Boeing Airplane Co.
ISBN 0-930403-09-6

***103. Thermophysical Aspects of Re-Entry Flows (1986)**
James N. Moss
NASA Langley Research Center
Carl D. Scott
NASA Johnson Space Center
ISBN 0-930403-10-X

***104. Tactical Missile Aerodynamics (1986)**
M. J. Hemsch, Editor
PRC Kentron, Inc.
J. N. Nielson, Editor
NASA Ames Research Center
ISBN 0-930403-13-4

***105. Dynamics of Reactive Systems Part I: Flames and Configurations; Part II: Modeling and Heterogeneous Combustion (1986)**
J. R. Bowen, Editor
University of Washington
J.-C. Leyer, Editor
Universite de Poitiers
R. I. Soloukhin, Editor
Institute of Heat and Mass Transfer, BSSR Academy of Sciences
ISBN 0-930403-14-2

***106. Dynamics of Explosions (1986)**
J. R. Bowen, Editor
University of Washington
J.-C. Leyer, Editor
Universite de Poitiers
R. I. Soloukhin, Editor
Institute of Heat and Mass Transfer, BSSR Academy of Sciences
ISBN 0-930403-15-0

***107. Spacecraft Dielectric Material Properties and Spacecraft Charging (1986)**
A. R. Frederickson
U.S. Air Force Rome Air Development Center
D. B. Cotts
SRI International
J. A. Wall
U.S. Air Force Rome Air Development Center
F. L. Bouquet
Jet Propulsion Laboratory, California Institute of Technology
ISBN 0-930403-17-7

***108. Opportunities for Academic Research in a Low-Gravity Environment (1986)**
George A. Hazelrigg
National Science Foundation
Joseph M. Reynolds
Louisiana State University
ISBN 0-930403-18-5

***109. Gun Propulsion Technology (1988)**
Ludwig Stiefel
U.S. Army Armament Research, Development and Engineering Center
ISBN 0-930403-20-7

*Out of print.

***110. Commercial Opportunities in Space (1988)**
F. Shahrokhi, Editor
K. E. Harwell, Editor
University of Tennessee Space Institute
C. C. Chao, Editor
National Cheng Kung University
ISBN 0-930403-39-8

***111. Liquid-Metal Flows: Magnetohydrodynamics and Application (1988)**
Herman Branover, Editor
Michael Mond, Editor
Yeshajahu Unger, Editor
Ben-Gurion University of the Negev
ISBN 0-930403-43-6

***112. Current Trends in Turbulence Research (1988)**
Herman Branover, Editor
Michael Mond, Editor
Yeshajahu Unger, Editor
Ben-Gurion University of the Negev
ISBN 0-930403-44-4

***113. Dynamics of Reactive Systems Part I: Flames; Part II: Heterogeneous Combustion and Applications (1988)**
A. L. Kuhl, Editor
R&D Associates
J. R. Bowen, Editor
University of Washington
J.-C. Leyer, Editor
Universite de Poitiers
A. Borisov
USSR Academy of Sciences
ISBN 0-930403-46-0

***114. Dynamics of Explosions (1988)**
A. L. Kuhl, Editor
R & D Associates
J. R. Bowen, Editor
University of Washington
J.-C. Leyer, Editor
Universite de Poitiers
A. Borisov
USSR Academy of Sciences
ISBN 0-930403-47-9

***115. Machine Intelligence and Autonomy for Aerospace (1988)**
E. Heer
Heer Associates, Inc.
H. Lum
NASA Ames Research Center
ISBN 0-930403-48-7

***116. Rarefied Gas Dynamics: Space Related Studies (1989)**
E. P. Muntz, Editor
University of Southern California
D. P. Weaver, Editor
U.S. Air Force Astronautics Laboratory (AFSC)
D. H. Campbell, Editor
University of Dayton Research Institute
ISBN 0-930403-53-3

***117. Rarefied Gas Dynamics: Physical Phenomena (1989)**
E. P. Muntz, Editor
University of Southern California
D. P. Weaver, Editor
U.S. Air Force Astronautics Laboratory (AFSC)
D. H. Campbell, Editor
University of Dayton Research Institute
ISBN 0-930403-54-1

***118. Rarefied Gas Dynamics: Theoretical and Computational Techniques (1989)**
E. P. Muntz, Editor
University of Southern California
D. P. Weaver, Editor
U.S. Air Force Astronautics Laboratory (AFSC)
D. H. Campbell, Editor
University of Dayton Research Institute
ISBN 0-930403-55-X

119. Test and Evaluation of the Tactical Missile (1989)
Emil J. Eichblatt Jr., Editor
Pacific Missile Test Center
ISBN 0-930403-56-8

***120. Unsteady Transonic Aerodynamics (1989)**
David Nixon
Nielsen Engineering & Research, Inc.
ISBN 0-930403-52-5

121. Orbital Debris from Upper-Stage Breakup (1989)
Joseph P. Loftus Jr.
NASA Johnson Space Center
ISBN 0-930403-58-4

122. Thermal-Hydraulics for Space Power, Propulsion and Thermal Management System Design (1990)
William J. Krotiuk
General Electric Co.
ISBN 0-930403-64-9

---
*Out of print.

123. Viscous Drag Reduction in Boundary Layers (1990)
Dennis M. Bushnell
Jerry N. Hefner
NASA Langley Research Center
ISBN 0-930403-66-5

***124. Tactical and Strategic Missile Guidance (1990)**
Paul Zarchan
Charles Stark Draper Laboratory, Inc.
ISBN 0-930403-68-1

125. Applied Computational Aerodynamics (1990)
P. A. Henne, Editor
Douglas Aircraft Company
ISBN 0-930403-69-X

126. Space Commercialization: Launch Vehicles and Programs (1990)
F. Shahrokhi
University of Tennessee Space Institute
J. S. Greenberg
Princeton Synergetics Inc.
T. Al-Saud
Ministry of Defense and Aviation
Kingdom of Saudi Arabia
ISBN 0-930403-75-4

127. Space Commercialization: Platforms and Processing (1990)
F. Shahrokhi
University of Tennessee Space Institute
G. Hazelrigg
National Science Foundation
R. Bayuzick
Vanderbilt University
ISBN 0-930403-76-2

128. Space Commercialization: Satellite Technology (1990)
F. Shahrokhi
University of Tennessee Space Institute
N. Jasentuliyana
United Nations
N. Tarabzouni
King Abulaziz City for Science and Technology
ISBN 0-930403-77-0

***129. Mechanics and Control of Large Flexible Structures (1990)**
John L. Junkins
Texas A&M University
ISBN 0-930403-73-8

130. Low-Gravity Fluid Dynamics and Transport Phenomena (1990)
Jean N. Koster
Robert L. Sani
University of Colorado at Boulder
ISBN 0-930403-74-6

131. Dynamics of Deflagrations and Reactive Systems: Flames (1991)
A. L. Kuhl
Lawrence Livermore National Laboratory
J.-C. Leyer
Universite de Poitiers
A. A. Borisov
USSR Academy of Sciences
W. A. Sirignano
University of California
ISBN 0-930403-95-9

132. Dynamics of Deflagrations and Reactive Systems: Heterogeneous Combustion (1991)
A. L. Kuhl
Lawrence Livermore National Laboratory
J.-C. Leyer
Universite de Poitiers
A. A. Borisov
USSR Academy of Sciences
W. A. Sirignano
University of California
ISBN 0-930403-96-7

133. Dynamics of Detonations and Explosions: Detonations (1991)
A. L. Kuhl
Lawrence Livermore National Laboratory
J.-C. Leyer
Universite de Poitiers
A. A. Borisov
USSR Academy of Sciences
W. A. Sirignano
University of California
ISBN 0-930403-97-5

134. Dynamics of Detonations and Explosions: Explosion Phenomena (1991)
A. L. Kuhl
Lawrence Livermore National Laboratory
J.-C. Leyer
Universite de Poitiers
A. A. Borisov
USSR Academy of Sciences
W. A. Sirignano
University of California
ISBN 0-930403-98-3

*Out of print.

***135. Numerical Approaches to Combustion Modeling (1991)**
Elaine S. Oran
Jay P. Boris
Naval Research Laboratory
ISBN 1-56347-004-7

136. Aerospace Software Engineering (1991)
Christine Anderson, Editor
U.S. Air Force Wright Laboratory
Merlin Dorfman, Editor
Lockheed Missiles & Space Company, Inc.
ISBN 1-56347-005-5

137. High-Speed Flight Propulsion Systems (1991)
S. N. B. Murthy
Purdue University
E. T. Curran
Wright Laboratory
ISBN 1-56347-011-X

138. Propagation of Intensive Laser Radiation in Clouds (1992)
O. A. Volkovitsky
Yu. S. Sedenov
L. P. Semenov
Institute of Experimental Meteorology
ISBN 1-56347-020-9

139. Gun Muzzle Blast and Flash (1992)
Günter Klingenberg
Fraunhofer-Institut für Kurzzeitdynamik, Ernst-Mach-Institut
Joseph M. Heimerl
U.S. Army Ballistic Research Laboratory
ISBN 1-56347-012-8

***140. Thermal Structures and Materials for High-Speed Flight (1992)**
Earl. A. Thornton
University of Virginia
ISBN 1-56347-017-9

141. Tactical Missile Aerodynamics: General Topics (1992)
Michael J. Hemsch
Lockheed Engineering & Sciences Company
ISBN 1-56347-015-2

142. Tactical Missile Aerodynamics: Prediction Methodology (1992)
Michael R. Mendenhall
Nielsen Engineering & Research, Inc.
ISBN 1-56347-016-0

143. Nonsteady Burning and Combustion Stability of Solid Propellants (1992)
Luigi De Luca, Editor
Politecnico di Milano
Edward W. Price, Editor
Georgia Institute of Technology
Martin Summerfield, Editor
Princeton Combustion Research Laboratories, Inc.
ISBN 1-56347-014-4

144. Space Economics (1992)
Joel S. Greenberg, Editor
Princeton Synergetics, Inc.
Henry R. Hertzfeld, Editor
HRH Associates
ISBN 1-56347-042-X

145. Mars: Past, Present, and Future (1992)
E. Brian Pritchard
NASA Langley Research Center
ISBN 1-56347-043-8

146. Computational Nonlinear Mechanics in Aerospace Engineering (1992)
Satya N. Atluri
Georgia Institute of Technology
ISBN 1-56347-044-6

147. Modern Engineering for Design of Liquid-Propellant Rocket Engines (1992)
Dieter K. Huzel
David H. Huang
Rocketdyne Division of Rockwell International
ISBN 1-56347-013-6

148. Metallurgical Technologies, Energy Conversion, and Magneto-hydrodynamic Flows (1993)
Herman Branover
Yeshajahu Unger
Ben-Gurion University of the Negev
ISBN 1-56347-019-5

149. Advances in Turbulence Studies (1993)
Herman Branover
Yeshajahu Unger
Ben-Gurion University of the Negev
ISBN 1-56347-018-7

150. Structural Optimization: Status and Promise (1993)
Manohar P. Kamat
Georgia Institute of Technology
ISBN 1-56347-056-X

---
*Out of print.

151. Dynamics of Gaseous Combustion (1993)
A. L. Kuhl, Editor
Lawrence Livermore National Laboratory
J.-C. Leyer, Editor
Universite de Poitiers
A. A. Borisov, Editor
USSR Academy of Sciences
W. A. Sirignano
University of California
ISBN 1-56347-060-8

152. Dynamics of Heterogeneous Gaseous Combustion and Reacting Systems (1993)
A. L. Kuhl, Editor
Lawrence Livermore National Laboratory
J.-C. Leyer, Editor
Universite de Poitiers
A. A. Borisov, Editor
USSR Academy of Sciences
W. A. Sirignano, Editor
University of California
ISBN 1-56347-058-6

153. Dynamic Aspects of Detonations (1993)
A. L. Kuhl
Lawrence Livermore National Laboratory
J.-C. Leyer
Universite de Poitiers
A. A. Borisov
USSR Academy of Sciences
W. A. Sirignano
University of California
ISBN 1-56347-057-8

154. Dynamic Aspects of Explosion Phenomena (1993)
A. L. Kuhl
Lawrence Livermore National Laboratory
J.-C. Leyer
Universite de Poitiers
A. A. Borisov
USSR Academy of Sciences
W. A. Sirignano
University of California
ISBN 1-56347-059-4

155. Tactical Missile Warheads (1993)
Joseph Carleone, Editor
Aerojet General Corporation
ISBN 1-56347-067-5

156. Toward a Science of Command, Control, and Communications (1993)
Carl R. Jones
Naval Postgraduate School
ISBN 1-56347-068-3

***157. Tactical and Strategic Missile Guidance Second Edition (1994)**
Paul Zarchan
Charles Stark Draper Laboratory, Inc.
ISBN 1-56347-077-2

158. Rarefied Gas Dynamics: Experimental Techniques and Physical Systems (1994)
Bernie D. Shizgal, Editor
University of British Columbia
David P. Weaver, Editor
Phillips Laboratory
ISBN 1-56347-079-9

***159. Rarefied Gas Dynamics: Theory and Simulations (1994)**
Bernie D. Shizgal, Editor
University of British Columbia
David P. Weaver, Editor
Phillips Laboratory
ISBN 1-56347-080-2

160. Rarefied Gas Dynamics: Space Sciences and Engineering (1994)
Bernie D. Shizgal
University of British Columbia
David P. Weaver
Phillips Laboratory
ISBN 1-56347-081-0

161. Teleoperation and Robotics in Space (1994)
Steven B. Skaar
University of Notre Dame
Carl F. Ruoff
Jet Propulsion Laboratory, California Institute of Technology
ISBN 1-56347-095-0

162. Progress in Turbulence Research (1994)
Herman Branover, Editor
Yeshajahu Unger, Editor
Ben-Gurion University of the Negev
ISBN 1-56347-099-3

163. Global Positioning System: Theory and Applications, Volume I (1996)
Bradford W. Parkinson, Editor
Stanford University
James J. Spilker Jr., Editor
Stanford Telecom
Penina Axelrad, Associate Editor
University of Colorado
Per Enge, Associate Editor
Stanford University
ISBN 1-56347-107-8

164. Global Positioning System: Theory and Applications, Volume II (1996)
Bradford W. Parkinson, Editor
Stanford University
James J. Spilker Jr., Editor
Stanford Telecom
Penina Axelrad, Associate Editor
University of Colorado
Per Enge, Associate Editor
Stanford University
ISBN 1-56347-106-X

*Out of print.

165. Developments in High-Speed Vehicle Propulsion Systems (1996)
S. N. B. Murthy, Editor
Purdue University
E. T. Curran, Editor
Wright Laboratory
ISBN 1-56347-176-0

166. Recent Advances in Spray Combustion: Spray Atomization and Drop Burning Phenomena, Volume I (1996)
Kenneth K. Kuo
Pennsylvania State University
ISBN 1-56347-175-2

167. Fusion Energy in Space Propulsion (1995)
Terry Kammash
University of Michigan
ISBN 1-56347-184-1

168. Aerospace Thermal Structures and Materials for a New Era (1995)
Earl A. Thornton, Editor
University of Virginia
ISBN 1-56347-182-5

169. Liquid Rocket Engine Combustion Instability (1995)
Vigor Yang
William E. Anderson
Pennsylvania State University
ISBN 1-56347-183-3

170. Tactical Missile Propulsion (1996)
G. E. Jensen, Editor
United Technologies Corporation
David W. Netzer, Editor
Naval Postgraduate School
ISBN 1-56347-118-3

171. Recent Advances in Spray Combustion: Spray Combustion Measurements and Model Simulation, Volume II (1996)
Kenneth K. Kuo
Pennsylvania State University
ISBN 1-56347-181-7

172. Future Aeronautical and Space Systems (1997)
Ahmed K. Noor
NASA Langley Research Center
Samuel L. Venneri
NASA Headquarters
ISBN 1-56347-188-4

173. Advances in Combustion Science: In Honor of Ya. B. Zel'dovich (1997)
William A. Sirignano, Editor
University of California
Alexander G. Merzhanov, Editor
Russian Academy of Sciences
Luigi De Luca, Editor
Politecnico di Milano
ISBN 1-56347-178-7

174. Fundamentals of High Accuracy Inertial Navigation (1997)
Averil B. Chatfield
ISBN 1-56347-243-0

175. Liquid Propellant Gun Technology (1997)
Günter Klingenberg
Fraunhofer-Institut für Kurzzeitdynamik, Ernst-Mach-Institut
John D. Knapton
Walter F. Morrison
Gloria P. Wren
U.S. Army Research Laboratory
ISBN 1-56347-196-5

***176. Tactical and Strategic Missile Guidance Third Edition (1998)**
Paul Zarchan
Charles Stark Draper Laboratory, Inc.
ISBN 1-56347-254-6

177. Orbital and Celestial Mechanics (1998)
John P. Vinti
Gim J. Der, Editor
TRW
Nino L. Bonavito, Editor
NASA Goddard Space Flight Center
ISBN 1-56347-256-2

178. Some Engineering Applications in Random Vibrations and Random Structures (1998)
Giora Maymon
RAFAEL
ISBN 1-56347-258-9

179. Conventional Warhead Systems Physics and Engineering Design (1998)
Richard M. Lloyd
Raytheon Systems Company
ISBN 1-56347-255-4

180. Advances in Missile Guidance Theory (1998)
Joseph Z. Ben-Asher
Isaac Yaesh
Israel Military Industries—Advanced Systems Division
ISBN 1-56347-275-9

181. Satellite Thermal Control for Systems Engineers (1998)
Robert D. Karam
ISBN 1-56347-276-7

---
*Out of print.

182. Progress in Fluid Flow Research: Turbulence and Applied MHD (1998)
Yeshajahu Unger
Herman Branover
Ben-Gurion University of the Negev
ISBN 1-56347-284-8

183. Aviation Weather Surveillance Systems (1999)
Pravas R. Mahapatra
Indian Institute of Science
ISBN 1-56347-340-2

184. Flight Control Systems (2000)
Rodger W. Pratt, Editor
Loughborough University
ISBN 1-56347-404-2

185. Solid Propellant Chemistry, Combustion, and Motor Interior Ballistics (2000)
Vigor Yang, Editor
Pennsylvania State University
Thomas B. Brill, Editor
University of Delaware
Wu-Zhen Ren, Editor
China Ordnance Society
ISBN 1-56347-442-5

186. Approximate Methods for Weapons Aerodynamics (2000)
Frank G. Moore
ISBN 1-56347-399-2

187. Micropropulsion for Small Spacecraft (2000)
Michael M. Micci, Editor
Pennsylvania State University
Andrew D. Ketsdever, Editor
Air Force Research Laboratory, Edwards Air Force Base
ISBN 1-56347-448-4

188. Structures Technology for Future Aerospace Systems (2000)
Ahmed K. Noor, Editor
NASA Langley Research Center
ISBN 1-56347-384-4

189. Scramjet Propulsion (2000)
E. T. Curran, Editor
Department of the Air Force
S. N. B. Murthy, Editor
Purdue University
ISBN 1-56347-322-4

190. Fundamentals of Kalman Filtering: A Practical Approach (2000)
Paul Zarchan
Howard Musoff
Charles Stark Draper Laboratory, Inc.
ISBN 1-56347-455-7

191. Gossamer Spacecraft: Membrane and Inflatable Structures Technology for Space Applications (2001)
Christopher H. M. Jenkins, Editor
South Dakota School of Mines
ISBN 1-56347-403-4

192. Theater Ballistic Missile Defense (2001)
Ben-Zion Naveh, Editor
Azriel Lorber, Editor
Wales Ltd.
ISBN 1-56347-385-2

193. Air Transportation Systems Engineering (2001)
George L. Donohue, Editor
George Mason University
Andres Zellweger, Editor
Embry Riddle Aeronautical University
ISBN 1-56347-474-3

194. Physics of Direct Hit and Near Miss Warhead Technology (2001)
Richard M. Lloyd
Raytheon Electronics Systems
ISBN 1-56347-473-5

195. Fixed and Flapping Wing Aerodynamics for Micro Air Vehicle Applications (2001)
Thomas J. Mueller, Editor
University of Notre Dame
ISBN 1-56347-517-0

196. Physical and Chemical Processes in Gas Dynamics: Cross Sections and Rate Constants for Physical and Chemical Processes, Volume I (2002)
G. G. Chernyi
S. A. Losev
Moscow State University
S. O. Macharet
Princeton University
B. V. Potapkin
Kurchatov Institute
ISBN 1-56347-518-9

197. Physical and Chemical Processes in Gas Dynamics: Physical and Chemical Kinetics and Thermodynamics of Gases and Plasmas, Volume II
G. G. Chernyi
S. A. Losev
Moscow State University
S. O. Macharet
Princeton University
B. V. Potapkin
Kurchatov Institute
ISBN 1-56347-519-7

198. Advanced Hypersonic Test Facilities (2002)
Frank K. Lu, Editor
University of Texas at Arlington
Dan E. Marren, Editor
Arnold Engineering Development Center
ISBN 1-56347-541-3

*Out of print.

199. Tactical and Strategic Missile Guidance, Fourth Edition (2002)
Paul Zarchan
MIT Lincoln Laboratory
ISBN 1-56347-497-2

200. Liquid Rocket Thrust Chambers: Aspects of Modeling, Analysis, and Design
Vigor Yang, Editor
Pennsylvania State University
Mohammed Habiballah, Editor
ONERA
Michael Popp, Editor
Pratt & Whitney
James Hulka, Editor
Aerojet-General Corporation
ISBN 1-56347-223-6

201. Economic Principles Applied to Space Industry Decisions (2003)
Joel S. Greenberg
Princeton Synergetics, Inc.
ISBN 1-56347-607-X

202. Satellite Communications in the 21st Century: Trends and Technologies (2003)
Takashi Iida, Editor
Communications Research Laboratory
Joseph N. Pelton, Editor
George Washington University
Edward W. Ashford, Editor
SES GLOBAL
ISBN 1-56347-579-0

*Out of print.